WATER LANDS

WATER LANDS

A vision for the world's wetlands and their people

Fred Pearce and Jane Madgwick

CONTENTS

7 **FOREWORD**

8 **INTRODUCTION:** EVERLASTING SWAMPS

UPLANDS AND COLD LANDS 14

19 **RUOERGAI PLATEAU, CHINA:**
Rewetting bogs on the roof of the world

31 **PARAMO DE SUNTARBAN, COLOMBIA:**
The secrets of Andean grasslands

39 **LAKE LOKTAK, INDIA:**
How hydropower broke the mirror of Manipur

49 **RIFT VALLEY, ETHIOPIA:**
East African lakes drained for a Valentine rose

61 **MOSCOW PEATLANDS, RUSSIA:**
From peat mines to bog wilderness

69 **SIBERIAN MIRES:**
Permafrost thawing and carbon under threat

PREVIOUS PAGES: The River Niger in northern Mali sustains a wetland on the edge of the Sahara Desert known as the Inner Niger Delta, which is home to two million fishers, farmers, and herders.

RIVER FLOODPLAINS 78

83 **ENGLISH FENS AND THE AMERICAN MIDWEST:**
Land grabs in 'dismal swamps'

91 **PANTANAL, BRAZIL:**
A future for the world's largest wetland

111 **RUPUNUNI, GUYANA:**
How to map the glistening grasslands

117 **TONLE SAP, CAMBODIA:**
Reversing river is the beating heart of the Mekong

125 **LAKE CHAD, WEST AFRICA:**
Dams, dykes and an international refugee crisis

133 **INNER NIGER DELTA, MALI:**
Desert jewel on the brink

141 **SUDD SWAMP AND MESOPOTAMIAN MARSHES:**
Refuges in times of conflict

149 **RIVER RHINE, EUROPE:**
River 'rectification' replaced by 'making room for the river'

INLAND SEAS, SWAMPS AND SUMPS 158

- 163 **SALTON SEA, CALIFORNIA:**
  Celebrating artificial and accidental wetlands
- 173 **LAKES PRESPA AND OHRID, BALKANS:**
  Wetland truce brings wider peace
- 183 **ARAL SEA, CENTRAL ASIA:**
  What happens when a sea dies
- 189 **CUVETTE CENTRALE, CONGO:**
  Jungle swamps and a climate time bomb

COASTAL DELTAS AND LAGOONS 196

- 201 **RUFIJI DELTA, TANZANIA:**
  Restoring wetland rights for wise use
- 213 **MISSISSIPPI DELTA, LOUISIANA:**
  Ol' man river keeps on rolling
- 221 **KERALA, INDIA:**
  Crazy backwaters in 'God's own country'
- 229 **VENICE, ITALY:**
  When a coastal lagoon becomes an ocean bay
- 239 **ACEH, INDONESIA:**
  Mangroves to fight the next tsunami
- 249 **BAY OF BENGAL, SOUTH ASIA:**
  Absorbing cyclones in India and Bangladesh
- 259 **WADDEN SEA, NETHERLANDS:**
  Breaking dykes to stop the ocean
- 267 **FLORIDA EVERGLADES, USA:**
  Start of a blue carbon revolution
- 275 **JAVA, INDONESIA:**
  'God willing we plan to stay'
- 288 **CONCLUSION:** GROUNDSWELL

- 292 END NOTES
- 297 INDEX
- 302 ACKNOWLEDGEMENTS
- 303 CREDITS
- 304 AUTHORS

'Managing land and water together is going to be key to reversing land degradation worldwide. Wetlands are critical for people and ecosystems. This has become crystal clear in dryland areas, such as around Lake Chad and the Aral Sea. The collapse of these wetland ecosystems has adversely affected the land, biological diversity and the well-being of the people. *Water Lands* presents a compelling and urgent call to action by all.'

Ibrahim Thiaw
EXECUTIVE SECRETARY, UN CONVENTION TO COMBAT DESERTIFICATION

'*Water Lands* is exceptional because it shines a light on the importance of understanding how water systems have shaped nature, cultures and economies. Through stories and evidence from wetlands the world over, *Water Lands* points out how these relationships have changed and how problems can be resolved by empowering local people as central players in harnessing water and nature to secure a more resilient future.'

David Nabarro
SPECIAL ADVISOR TO THE UNITED NATIONS SECRETARY GENERAL ON THE 2030 AGENDA
FOR SUSTAINABLE DEVELOPMENT AND CLIMATE CHANGE

'Wetlands are the vital organs of a living planet. They nurse life, stabilize the climate, and anchor the water cycle locally and globally. But, as *Water Lands* makes plain wetlands are precious beyond carbon storage or biodiversity improvement. The examples show us the possibility that we might reverse the destruction of wetlands and of ecosystems generally. That destruction is as old as civilization. Might we forge a different kind of civilization embodying reverence for soil and forests, animals and plants, water and land – and the wetlands where they all meet?'

Charles Eisenstein
AUTHOR AND SPEAKER

'*Water Lands* is a timely intervention and should stir people into action. In particular, to find more equitable ways of sharing water. Each chapter directs attention to the much-needed re-orientation between rapid economic development and long-term prosperity. Wetland communities, among other indigenous people and the ecosystems they depend on, need more support in conserving the environment for their long-term prosperity and peace.'

Ikal Ang'elei
DIRECTOR, FRIENDS OF LAKE TURKANA

FOREWORD
WATER LANDS

WHERE WATER MEETS LAND, LIFE ABOUNDS. From the tundra mires to the temperate swamps, tropical forested rivers, desert oases, and deltas, it's the dynamic interaction of water and land that matters. Wetlands adorn, connect, and permeate our landscapes, storing and regulating water and providing vital stepping stones for migratory water birds.

From my early days as an ecologist, these magical ecosystems were the obvious choice for my focus and conservation efforts. But traditional conservation approaches have proven inadequate to save them. I have witnessed decades of draining, taming, conversion, and fragmentation and now see wetland natural resources being hotly contested, on the frontline as humans demand more water, food, and energy. In many places, water scarcity as well as rising floods are placing a limit on development and endangering peace. As risks related to devastating floods increase, the benefits of investing in wetlands as climate buffers, and as a basis for food security and equitable development, are increasingly recognized. But in terms of global wetland recovery, we are still settling into the starting blocks. Science is an insufficient basis for driving action. Accelerating the necessary change will depend on people revering and valuing wetlands more.

After meeting Fred Pearce some years ago at the Smithsonian Institution in Washington DC, following his impassioned speech about the tragedy of the Aral Sea and other wetlands of the world, I knew instantly that we should work together. His storytelling reportage has since helped Wetlands International explain to a wider audience how enabling communities to better manage wetlands can improve their well-being and build resilience across whole landscapes. Over recent years, we have discussed the need for a book which would help to lift and refresh the image of wetlands and serve as a call to action.

With *Water Lands,* we bring you stories, images and examples from wetlands around the world which reveal what is really happening and what is at stake in future choices. We aim to instil a sense of urgency and also hope. Most of all, we aim to inspire you to follow, support, and join efforts that will result in a turnaround for wetlands, their nature, and peoples. ○

Jane Madgwick
CHIEF EXECUTIVE OFFICER, WETLANDS INTERNATIONAL

INTRODUCTION
EVERLASTING SWAMPS

'EVERYTHING HERE DEPENDS on the water, but the government is taking our water. They are giving it to foreign farmers. The lakes don't fill properly now.' Daouda Sanankoua, Mayor of Deboye in Mali, was describing the plight of his constituents in the heart of the Inner Niger Delta, on the edge of the Sahara Desert. Their water, collecting in a wetland the size of Belgium, once nurtured a great civilization around the fabled city of Timbuktu. Today, two million people depend on its fish, and its waters washing across croplands and pastures. Millions of European birds migrate to the delta each winter. But dams upstream are diverting water away from the delta. 'The wind is driving sand into our village,' said one of Sanankoua's people. 'Most of our fields are gone.'[1]

With another dam awaiting construction, hydrologists say that soon every year on the delta could be a drought year. Yet the people here are fighting the encroaching desert. They are planting trees and grasses, channelling water to fields, creating vegetable gardens, and digging fish ponds. Not a drop of water must go to waste, they say. They want the dams upstream to be managed to allow the delta to continue to flood each wet season. Without the annual flood, the prognosis for the wetland, its people and its birdlife, is bad. As Sanankoua put it, 'a wetland is nothing without its water.'

From the peat bogs of Ireland to the bayous of Louisiana; from the flooded forests of Cambodia to the permafrost of Siberia; from the mangroves of the Ganges Delta to the 'everlasting swamps' of the Nile; and from the marshes of the Brazilian Pantanal to the boggy upland pastures of Tibet, wetlands are in-between and ever-changing worlds. Sometimes wet and sometimes dry, sometimes land and sometimes water, sometimes saline and sometimes fresh; they change character with the seasons, or may lie dormant for decades before bursting into life.

The Bible says that God created the world by dividing the land from the water. If so, He forgot about wetlands. For in wetlands rivers have no fixed banks, water oozes through soil, and silty soil courses through water. Fish live in trees and

The Garden of Eden, a Christian image of paradise, may have been based on the Mesopotamian Marshes in modern-day Iraq.

land mammals swim for their lives. Yet despite their ephemeral and contradictory nature, wetlands are our planet's richest natural resource. Wetlands have extraordinary natural abundance. And that abundance helps make our world. Fertile with silt and rich in wildlife, wetlands were the first and most important sources of wealth for early human civilizations. From ancient Egypt to Angkor Wat in Cambodia, and Mesopotamia to Timbuktu, our earliest urban societies grew on and beside wetlands.[2]

Wetlands regulate and govern the hydrology of most of the world's river systems and determine access to water for billions of people. They nurture fisheries and water crops, maintain dry-season pastures, and provide fresh water and flood protection far downstream. They make deserts liveable by cooling summers, warming winters, and wetting dry air. They maintain river flows and top up groundwaters. They protect shorelines from high tides and storms. They simultaneously provide irrigation and drainage. They are the planet's biggest terrestrial carbon store. They are home to a third of all known vertebrate species.[3] They even reduce pollution by soaking up nutrients and toxins.

Finally, many wetlands are among the last great areas of the planet held in common by their communities. Their unfenced wealth is open to nomads in search of pasture for their animals, but also to the poor, the distressed, the marginalized and the banished. In war-torn and drought-stricken lands, they may provide security and the next meal for those *in extremis*.

WETLANDS CONJURE UP MANY IMAGES: lazy lagoons, gardens of Eden, mysterious mist-shrouded Shangri Las, or shimmering lakes. The Ramsar Convention on Wetlands came up with forty-two types.[4] But categorizing them is a thankless task, and this book won't get 'bogged down' in definitions. Instead, as we journey from source to sea in search of their past, present, and future, we will revel in their sheer variety and the range of services they provide for nature and for us.

All are rich in wildlife, of course. Insects, birds, and fish especially. Seemingly isolated wetlands across the world are parts of a network of watery places that sustain migrating water birds. Travelling from North to South America; from Europe to Africa; across Asia; and from the poles to the tropics, birds depend on these vital stepping stones to rest, feed, and breed. Remove one wetland and you may decimate bird populations thousands of kilometres away.

But wetlands provide myriad human services too. We will visit many peat bogs, for instance. These large waterlogged stores of plant carbon occur everywhere from mountain valleys to coastal swamps, and from Arctic tundra to tropical rainforests. They help protect the planet from the worst of climate change, but if drained or burned they can rapidly become part of the climate problem. We will travel to dry lands, such as the African Sahel, where the precarious fate of the Inner Niger Delta, Lake Chad, and the Sudd swamp on the Nile is a matter of life and death for tens of millions of people. On coasts, we will see how the salty waters of river deltas, estuaries, mangrove swamps, mudflats, lagoons, and salt marshes, sustain fisheries and water crops, and defend hundreds of millions of people from the ravages of the oceans.

Wetlands are also part of the great river systems that flow from mountain peaks to the oceans. In the modern world, we too often think of rivers as water conduits cut off from the surrounding land. But natural rivers meander and divide, stop off in lakes and marshes, form deltas and spread across verdant floodplains. Most rivers retain at least some of these features. Rivers are wetlands in their own right.

Canada's mighty River Mackenzie connects no fewer than three large deltas that punctuate its path to the Arctic Ocean, and a huge floodplain known as the Mackenzie Lowlands. In Asia, the Ganges surges from glacial lakes in the Himalayas, through the swamps and flooded grasslands of the Terai and its wide floodplain – one of the most densely populated regions in the world – to the Sundarbans, the world's largest mangrove swamp. The upper reaches of the River Paraguay in South America pass through the Pantanal, the world's largest wetland, before flowing through others almost as large to the Paraná Delta. In Central Europe, the Danube passes through wetlands left undisturbed by the Cold War on its journey to its own delta, Europe's largest, and the Black Sea.

Wetlands have few fixed boundaries. They change as the water comes and goes. So global estimates of their total extent vary widely. But at any one time, they certainly exceed the size of the United States or China. In the recent past, these watery commons have been Cinderella ecosystems, shunned or forgotten. In European folklore and literature, swamps, mires, bogs and moors have been places to be feared, where monsters roam and diseases lurk. 'As you value your life or your reason, keep away from the moor,' said Sir Charles Baskerville in Arthur Conan Doyle's Sherlock Holmes novel, *The Hound of the Baskervilles*. In British criminal history, few episodes loom as large as the 1960s child slayings known as the 'Moors murders', because of the children's bleak final burial ground.

No wonder that we have long sought to banish wetlands from our landscapes. 'Drain the swamp' is an old cry. Ever since the Dutch began installing windmills to pump out the boggy places where the Rhine met the North Sea, engineers have been on the case. During the twentieth century, around two-thirds of all the world's wetlands – swamps, bogs, river banks, lakes, mires, billabongs, oases, deltas, mangroves, lagoons, ponds, mudflats, fens, floodplains and the rest – were drained, dammed and dyked to death. They are still disappearing three times faster than rainforests, but with far less protest.[5]

In their places are cities, infrastructure and farmland – all released from the menace of flooding. Or so it is claimed. For the protection is often illusory. We now increasingly realize that wetlands prevent more floods than they cause. And rather than being badlands, they are a good thing, nourishing landscapes and maintaining river systems. Just as the jungles of past nightmares are now rainforests to be cherished, so the watery wastes are being reimagined as places teeming with natural wealth, vital to biodiversity and human welfare, and helping to resist climate change. As the first President Bush declared as long ago as 1989, 'It's time to stand the history of wetlands destruction on its head; from this year forward, anyone who tries to drain the swamp is going to be up to his ears in alligators.'[6]

Our new-found appreciation of wetlands is tapping into old roots. In much of the world, wetlands have always been important cultural beacons. Many are sacred places. Lake Loktak in north-east India is the spiritual home of the Meitei people.[7] The Wapichan in Guyana chart their ancestry in their flooded grasslands. For Christians, the founding story of humanity begins in the Garden of Eden, widely regarded as the Mesopotamian Marshes of modern-day Iraq. And even where that religious association is lost, we have always found beauty in their open spaces: whether the Lake District in England, southern Africa's Okavango Delta, India's Kerala Backwaters, Peru's Lake Titicaca or the Florida Everglades.

But the wealth of wetlands remains, literally and metaphorically, hard to grasp. They could yet slip through our fingers. The fluid, unfenced and often unowned attributes of wetlands are difficult to reconcile with a modern world that values certainty in nature more than variability, where floods are seen primarily as destructive, and where private property is the universal currency. Despairing of exploiting the moving feast, our first instinct too often remains to enclose, privatize, and tame; to dam, drain, dyke, dredge, canalize, or concrete them over.

This is madness. As one assessment by a collaboration of international scientists concluded, 'our relentless conversion and degradation of remaining natural habitats – including wetlands – is eroding overall human welfare for short-term private gain'.[8] It is simply a myth to imagine that a degraded environment is the inevitable price of development. Quite the contrary. In the Anthropocene, we need wetlands and the wealth they provide at least as much as we need rainforests or healthy oceans.

So, besides charting the losses and exposing the continuing threats that wetlands face in the twenty-first century, the central purpose of this book is to explore and celebrate ways that the world's wetlands can and must be secured and restored, for a healthy planet and for healthy and well-fed humans. The stories told here are at least as much about wetlanders as wetlands.

Many conservationists would like to keep wetlands for nature alone. Especially for their birdlife, which happily also commands high value among tourists. In places, that remains possible. But wetlands are often too important to the people that live in and beside them to be set aside for nature alone. They provide sustenance for millions, and refuges and security for many poor and marginalized communities. They maintain river systems on which billions, often far downstream, depend. Moreover, as shared and commonly owned resources, their protection and restoration can sometimes help nurture peace – between communities and even between nations, such as in the Balkans, as we shall see.

The challenge today is to recognize the vast natural capital tied up in wetlands, and to access that capital by reinventing wetlands for the modern world, with new systems of governance, new economic roles, new crop regimes, and, if necessary, new hydrology. It means managing rivers and lakes to restore natural flows and reconnect them with their floodplains. In many places, it also means taking advantage of entirely artificial and accidental wetlands that have sprung up across the landscape, and to recognize that urban wetlands are no longer an oxymoron.

Wisely used, wetlands help development and lift people out of poverty. Indeed, this can be the key to their conservation. For if we don't value wetlands, if they don't contribute to the local economy and the livelihoods of their inhabitants, then they will be forsaken and lost.

Of course, a commitment to 'wise use' raises as many questions as it answers. Does it mean allowing in a few fishers and reed cutters? Or how about cattle herders? And if them, then why not farmers and charcoal burners? Should the arbiters of what is meant by wise use be economists, weighing farm profits against a dollar value attached to ecosystem services? Or ecologists, who may have a very different way of measuring wetland wealth? Or how about the wetland residents themselves, and the millions who can benefit from silt supplies and flood protection all the way to the ocean? No answer will fit all circumstances, but many people will conclude from the examples in this book that sustainable wetland use will only be achieved when those who live in and around them, and depend on their resources, have the final say.

What is certainly true is that if we want water for all we must take care of the wetlands that are the sources of most of our water, and rediscover ways to capture and store water in our landscapes. If we want safe coastlines, we need to revive the deltas and lagoons, mangrove swamps, and salt marshes that are nature's flood defences. If we want to reverse land degradation, restore natural capital and improve the lives of wetland inhabitants, then we need to block up drains, tear down dykes, and guarantee that free-flowing rivers can deliver their life-giving flood pulses. Wetlands matter. ○

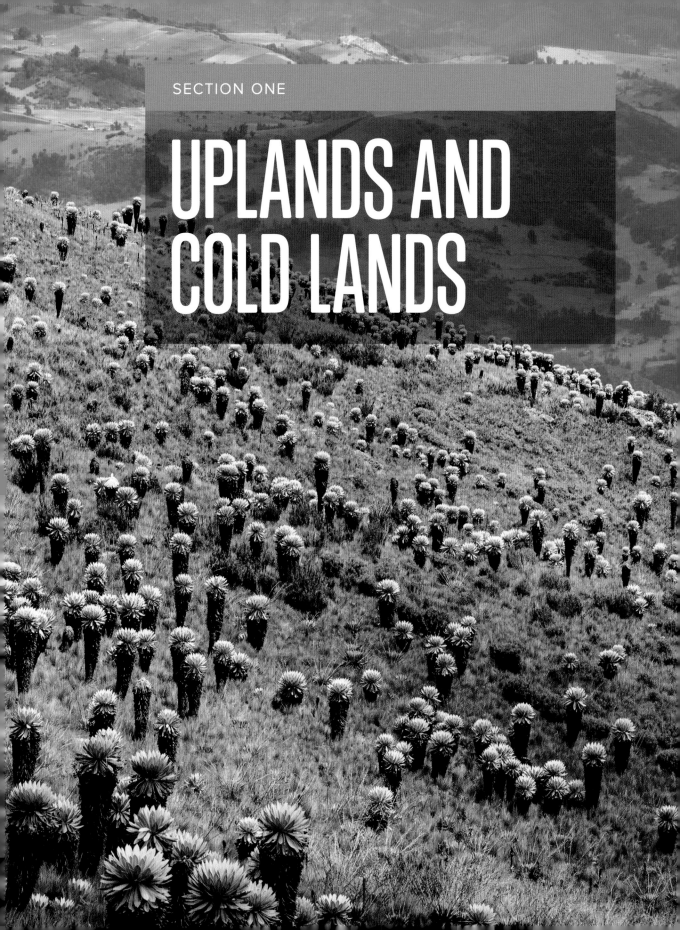

SECTION ONE

UPLANDS AND COLD LANDS

From the edges of Tibet to the mist-shrouded mountains of the Andes and the valleys of north-east India, upland bogs, lakes, and springs are where rivers begin. These wetlands regulate river flows for thousands of kilometres downstream. Some have been drained; others are locked up in permafrost now at risk of thawing. But many are also the sites of some of the largest wetland restoration projects in the world. Our journey begins with one such project among the yak-grazed bogs at the headwaters of the Yellow River in China.

PARAMO DE SUNTARBAN, COLOMBIA

■
PREVIOUS PAGES: High in the Andes Mountains, above the trees but below the glaciers, lie waterlogged and frequently misty worlds that supply Colombia with most of its water. The *páramos* are botanical El Dorados, rich in *frailejones*, a type of shrub whose long, spiky leaves are covered in tiny hairs that capture moisture from the air.

LOCATIONS FEATURED IN SECTION ONE

UPLANDS AND COLD LANDS 17

RUOERGAI PLATEAU, CHINA
REWETTING BOGS ON THE ROOF OF THE WORLD

Over 3,400 metres up on the edge of Tibet in central China, the air was thin and bitterly cold. It was −5° Celsius in the midday sun. The grasses of the Ruoergai Plateau stretched for more than 50 kilometres in every direction. In summer the plateau would have been boggy, but this was January, and it was frozen hard. With little snow, it was far from empty, however. More than a thousand yaks were visible with a sweep of the eye. That was far too many. The ancient cattle were grazing at densities more than twice the capacity of the land. The remote upland wilderness was under pressure, with water tables falling, grass drying, and dust kicked up by the animals' hooves filling the air.

China's Ruoergai Plateau, often called by its Tibetan name of the Zoige Plateau, covers 27,000 square kilometres, an area larger than Wales. It is a land of long, bitter winters, when the bogs and lakes are frozen to a depth of more than a metre, and short, brilliant summers, when everything melts and the grassland is awash with water. Humans have been here for 5,000 years, grazing herds of yaks on the abundant pastures. 'The yak made permanent occupation of the harsh environment of the plateau possible,' says Hans Joosten of Greifswald University in Germany.[9] The hairy and hardy animals prospered in the inhospitable terrain, providing Tibetan herders with milk, meat, hides, hair, transport, and dung for fuel.

On the face of it, not a great deal has changed. The yak remains central to the lives of most people today. The plateau is dotted with herders' meagre homesteads, around which the yaks graze during the winter, before moving to the surrounding mountains in summer. Women can often be seen outside the homesteads gathering the animals' dung to cook and heat their homes. But there have never been so many yaks up here. Overgrazing is rife, say Chinese officials. Huddled in overcoats in draughty offices, they talk openly of the 'desertification' of the largest high-altitude wetlands in the world. But they also describe their efforts to turn the tide, by rewetting dried-out areas and planting shrubs and grass to halt the desert's advance. So, is this a tragedy unfolding on the roof of the world, or the start of a new beginning? We had come to find out.

The heart of the matter is water. A sixth of the Ruoergai Plateau is made up of peat. In places it is more than 10 metres thick. The peat holds huge amounts of water, forming lakes and making the land so boggy in summer that it is often impossible to walk across. The plateau is like a giant reservoir. It sits at the head of two of China's biggest rivers, the Yangtze and the Yellow River. In particular, it is the largest source of water for the Yellow River, delivering almost 5 cubic kilometres each year into a national artery that has irrigated crops across northern China for thousands of years.[10] Hundreds of millions of Chinese depend on it. In the dry season, 45 per cent of the water in the river's upper reaches comes from the plateau's peat.[11] Or it did. For in the past half century, the sponge has been squeezed by efforts to modernize traditional Tibetan yak herding.

In the 1960s and 1970s, at the height of his Cultural Revolution, Mao Zedong sent cadres of Red Guards to the grassland. They dug more than 700 kilometres of drains into the boggy plateau.[12] The aim was to dry out the wetter pastures, so the yaks could graze year-round without drowning. There would be more milk and meat for a hungry country with a fast-growing population. Herders were encouraged to grow their herds. Livestock numbers almost quadrupled. Yak numbers in Ruoergai county, the plateau's main administrative district, rose from 112,000 to 510,000 between 1959 and 2006.[13]

A black-necked crane flies in to Lake Hua where the birds breed among the sedge marshes.

PAGE 18:
The Ruoergai Plateau on the edge of Tibet is one of the world's largest high-altitude peatlands. It feeds water to China's two greatest waterways, the Yangtze and Yellow rivers. Lake Hua, home to the rare and revered black-necked crane, has doubled in size following to restoration programme to protect the peat.

But as the drains lowered water levels across almost 50 per cent of the plateau's peatlands, the end result was not better pastures. Water rushing into the drains led to unprecedented erosion.[14] 'Key [water] channels have been eroded as deep as two metres,' worried investigators from the Ramsar Convention on Wetlands reported. A cascade of environmental degradation followed. Of the plateau's seventeen lakes, six dried out entirely, while the other eleven all shrank.[15] The desiccated peat soils proved a great home for small burrowing mammals such as the Himalayan marmot, the rabbit-like

pika, and the *zokor*, which resembles a mole rat. Rodent densities reached 300 per hectare in places, worsening the damage by grazing and digging holes in the peat.[16]

The grasslands weren't pristine before. But according to government data, the degraded area rose by 50 per cent. It included three-quarters of the 4,700 square kilometres of peatland. A once-verdant wetland was in places turning to a dustbowl. This was bad for the yaks and bad for the nation's hydrology. The sponge that had provided a constant supply of water into the Yellow River now delivered more peak flow but less of the critical dry-season flow.

Combined with declining rainfall and increased abstraction for irrigation downstream, the Yellow River began to dry up. Since 1972, that has happened more than thirty times, sometimes as far as 550 kilometres from the ocean.

With hydrological disaster looming, and following an intervention by Wetlands International, in 1999 the government banned further drainage of the peatlands. Four years later, after mains electricity arrived in the region in 2003, it also halted the once-widespread cutting of peat for fuel. It created five nature reserves, including the Sichuan Ruoergai Wetland National Nature Reserve, which covers 116 square kilometres in the heart of the plateau. And from 2004, local authorities have collaborated with Wetlands International and Tibetan herders to rewet parts of the drained marshes.

Since then, hundreds of dams have been inserted into the drains. Most were small, made with peat, wooden planks, or sandbags. But bigger drains required stone and concrete. The raised water levels in the ditches has enabled marsh plants to recover and the restoration of lush marshes have simultaneously improved summer grazing for yaks.

It is ongoing work but, so far, some 64 square kilometres of plateau wetlands have been restored, says Gu Haijun of Sichuan Wetlands Management Center in Chengdu, who has been involved in the task for twenty years. It is one of the world's largest wetland restoration projects – and one that few people outside this remote region have heard of.

Our tour of the plateau found many successfully restored areas of marsh. The small dams were often all but submerged by accumulating peat and knee-high vegetation in restored gullies. On a bigger scale, blocking drains around the plateau's largest water body, Lake Hua, had doubled its surface area to 6.9 square kilometres.[11] The recovery of water levels in the lake was proving a boon to wildlife in an area increasingly seen in China as a summer eco-tourist attraction.

After taking a precarious walk out onto the frozen Lake Hua, we met Ruke, a trim and shy Tibetan who had worked for eleven years superintending the lake, which sits inside the Ruoergai Nature Reserve. He said the revived lake has seen a resurgence of birdlife. Most of the 200-plus bird species that visit the reserve in summer congregate near the lake. Even in mid-winter there were swans circling outside. Ruke's main charge and delight is the black-necked crane, a bird revered by Tibetans. A tenth of the world population, around a thousand pairs, fly in to breed on the Ruoergai Plateau each spring. Ruke was proud that in the previous March he had been the first person to spot their return. 'I heard them sing during the night. I contacted people and it was on TV the next day.'

PREVIOUS PAGES:
Half a century ago, cadres of Red Guards dug more than 700 kilometres of drains into the boggy plateau with the aim of improving yak pastures. But the result was worse pastures and spreading desert. Now the drains are being plugged to restore the wetland.

He knows the cranes' habits intimately. They nest mostly in sedge grass close to the lake. 'Some raise their young right by my home here,' he said. 'I can go close and they don't fly away. I think they know I am not a threat. We love them. For Tibetans, they are a sign of good luck.' Recently, they had taken to nesting close to a boardwalk across the lake, a favourite attraction for tourists. Ruke was happy that his bosses running the reserve had removed the boardwalk to protect the cranes. Its steel remains were piled up on the shore of the frozen lake.

Nothing, not even the tourist industry, is allowed to get in the way of protecting the cranes. They have become the symbol of conservation in the area. Buildings in local towns we visited had almost as many images of the birds as they did of the more traditional local icon of nature's bounty, yak horns. They have become a symbol too of the successful rewetting of the Ruoergai Plateau.

The full restoration of the Ruoergai grasslands will require more than plugged drains, however. Despite the improvements to their grazing grounds, everyone agrees that there are still too many yaks. But how many is too many? 'It depends who you ask,' said Gu. While some people see yaks as part of the natural landscape, requiring only a modest reduction in numbers, others regard them as a scourge that should be eliminated.

The debate hangs, in part, on the role of the animals in the plateau's ecological history. Wetland specialists such as Joosten say that, long before the drains were dug, the yaks changed everything on the plateau. The constant trampling of their hooves converted most of what was once a 'percolation mire', in which water freely seeped down through the bog, into the 'surface-flow mire' that dominates today, in which the peat surface is compacted and the water ponds up in the rich bog vegetation.

Joosten does not recommend banishing the yak. It is too late for that. 'Thousands of years of grazing have created a new landscape of surface-flow mires that is recognized as a beautiful natural and cultural heritage.' It is 'one of the most impressive open landscapes in the world,' he says. 'Although the origin of these surface-flow mires must be considered to be degradation itself, it cannot be the aim to restore the former percolation mires over large areas.' It would be physically impossible, 'because over large areas the peat hydraulic properties have changed irreversibly'.[12] Compaction is forever.

So, what should happen? What density of yaks would prevent further degradation? They have been considering this issue at the nature reserve, on whose grasslands an estimated 150,000 livestock are allowed to graze during the winter. That is one animal for every 0.08 hectares. Not all are yaks. There are many sheep, and a few horses and cattle. But according to Erga, Director of the reserve, each yak needs 1.2 hectares of grazing and each sheep 0.4 hectares. So, even though the animals only graze on the plateau for half of the year, their numbers are clearly well in excess of its carrying capacity.

Conservation is a priority in China, these days. On our visit, many people repeated President Xi's call for the country to create an 'ecological civilization'.[17] Many would like the yaks removed on principle, to let the peatland recover. 'The original idea at the reserve was to totally stop grazing, in winter as well as summer, to protect its biodiversity,' said Yong-Xiu, Deputy Director of the reserve. But recent research conducted at the reserve suggests that is not such an eco-friendly option – as we discovered when we joined Yong-Xiu on a visit to a 200-hectare experimental site within the nature reserve, where all grazing had ended three years previously. It is one of a number of sites where they are testing out different models for future management of the reserve.

The fenced-off area looked magnificent. Even in the midst of winter, tall grasses such as Dahurian wild rye were everywhere. It was a sharp contrast to the stubby grass in areas outside the fence. But the preliminary findings of the experiment are that the magnificence was not all that it seemed. The long grasses may look good, but they are out-competing other species. Some herbs lose out, and raptors are losing habitat. So are the black-necked cranes, which prefer the lower, grazed sedges for their nesting grounds. In the current situation, keeping yaks off the grasslands would probably reduce biodiversity, according to Yong-Xiu. 'We need the yaks to maintain the biodiversity.'

That message will be good news for the people who currently live and graze their yaks there, including those we met at Xiangdong, an old settlement deep within the reserve. Despite their remote location they have water, electricity, a clinic, a veterinary practice, a kindergarten and a shop. We saw sheds stuffed with yak dung to keep the stoves going through the winter. The streets have solar-powered lights, there is a basketball court and playground, a small temple, and a noticeboard featuring the day's newspapers.

But all this depends on their being able to carry on grazing their yaks. We met Luorangnimei, the village head, huddled round a stove in the community meeting room. Luorangnimei, also the village vet, told us that each of its 136 households has around seventy yaks and sheep. The government has restricted the size of their herds, but pays them a subsidy in return. 'We have to balance livelihood and conservation,' he said. 'We realize that livestock numbers will be reduced further in future.'

Yak grazing remains a way of life for most families on the plateau. But the government is introducing new economic activities such as tourism and is paying herders to become nature wardens.

But they could not imagine being without their herds. Yong-Xiu, herself a Tibetan, said that whatever the ecological issues, 'the yak is life for the Tibetans. It has always provided everything. The yak horn is a symbol of our identity and is revered in our Buddhist religion. Yaks bring status. They are still seen as an indication of a person's wealth.'

So, while there is a clear need to reduce yak numbers on the plateau, there is a softly-softly approach to winding down Tibetan yak culture. Nobody wants to pick a fight with the Tibetan communities. Different localities are coming up with their own solutions. Not far from the reserve, in Hongyuan county, Vice-Governor Liu Jian told us that families there were restricted to eight yaks per person, less than half the twenty-two reckoned necessary for a decent livelihood. To make up the difference, the county offers subsidies for giving up herding lands, and encourages herders to make extra cash by tapping into the growing local tourist trade.

During our visit Hongyuan was almost deserted. Its restaurants, shops, and hotels were mostly closed for the winter. But in summer Chinese tourists come in increasing numbers. It has a visitor centre and tourist villages. There is a yak music festival each August. The area was an important stopping-off point during the Long March of Communist revolutionaries back in 1935. Visitors to the nearby Waqie Wetland can take a boardwalk over the bog signposted as the 'Red Army's Long March Experience Avenue'. Over the road, the county provides stalls where herders sell the tourists anything from yak milk to horse rides.

If our experience is anything to go by, sampling Tibetan herders' domestic hospitality should be a draw, too. After we toured Lake Hua, Ruke took us to his tiny one-roomed house by the lake, where he lives with his wife, pigtailed Lamujian. She is in charge of their herd of fifty yaks. She stays at home with her husband in winter, while the animals graze on the reserve's frozen grasslands. But in summer, she takes them to pastures in the mountains, where she lives a nomadic life, sleeping in a tent. We huddled close to their dung-burning stove as Lamujian prepared a traditional Tibetan lunch. It began with a starter of rice, yak butter and the rhizome of potentilla, a herb which grows locally. The main course was dumplings stuffed with local yak meat. We finished with *tsampa* – a snack made of barley flour, tea, yak butter and sugar, which she mixed for us in a bowl. Such a life must be as novel for China's half-billion urbanites as for most Europeans. What a tourist experience for them.

The authorities on the plateau have a second way of weaning herders off their dependence on yaks. They want to turn some of them into conservation rangers. Hongyuan county employs more than a thousand poor herders to scour the grasslands for intruders, maintain fences and look after wildlife. The nature reserve is doing the same thing. We met one of them, who neatly combined the two worlds here.

Shangte is every inch a man of the plateau, wrapped in traditional Tibetan garb against the cold, including double-length jacket sleeves that dangle almost to the ground but keep hands warm without gloves. Despite his new day-job, he keeps a herd of fifty yaks himself. Two of his three children are also herders. But he has another side. He talked of his love of the wildlife that he is now in charge of: the

Blocking drainage channels constructed half a century ago across the peatland allows water levels to recover. This is one of the world's largest peatland restoration projects.

Tibetan gazelles, young wolves playing in the grass, and the six pairs of cranes down by the stream that he keeps a close eye on.

There is no going back to the old days on the plateau. Herders today tend their animals from motorbikes rather than horses, and mostly live in houses on the edge of towns rather than in tents. Their children attend schools and may go to university, like Shangte's third child who is studying traditional Tibetan medicine at Lhasa University. And market forces are part of their daily lives. Yaks are profitable. A healthy four-year-old animal going for slaughter fetches around $750. There

is a rising demand in Chinese cities for lean yak meat, as well as dairy products such as yoghurt. They are seen as green and organic.

Yong-Xiu – from her position as both a Tibetan and the Deputy Director of the nature reserve – is confident that Tibetan herders could make the most of their commercial opportunities without squandering their environment. She believes the way to restore the peatlands is to sustain Tibetan culture and traditions, rather than dismembering them. Is that over-optimistic, given the current overgrazing? She thinks not. 'We need a balance between conservation and livelihoods, and the herders know best how to achieve that balance.'

Herders may be underlining her faith in them in innovative ways, says Yale anthropologist, Gao Yufang, who has been researching rangeland management in the region. He told a story about the rise and fall of fencing on the plateau. When the government broke up the country's farming collectives at the end of the twentieth century, it privatized the Ruoergai grasslands. It issued its yak herders with long-term grazing rights to individual parcels of land. The herders naturally fenced off their land to keep other people's animals away. As a result, most of the plateau is now crossed by fences.

The central government continues to push for extensions to private land ownership and has targets for erecting more fences on the plateau, as elsewhere in China.[18] But that is proving bad news for the grassland. Driving through the area, we saw vivid contrasts either side of the fences. One side was denuded of grass due to being grazed by a large herd, while metres away over the fence an area of lush grass was ungrazed. It makes no sense for either the environment or the herders.

In the old days, the herders would have moved their herds around to find the good grass and avoid the bad. But the fences prevent that, says Gao. And the herders recognize the problem. So, they are adapting the system themselves, by pooling the herds of different households and tearing down the fences between them. We met several herders who said they were doing just that. It seems a promising way forward, and some government officials are now giving their support, Gao said.

As we left the plateau, the bitter Tibetan wind was intensifying. Near Tangke, a holy site overlooking the Yellow River, yaks were kicking up dust that caught in the wind, while close by others were slipping on ice in search of grasses. It was a strange contrast: a land where abundant water and the spread of deserts persist side by side.

This land has evidently been dramatically changed by yaks, over many thousands of years. There is much to do. But outsiders should be wary of intervening too strongly. Such interventions have gone wrong before. Perhaps only the Tibetan yak herders hold the key to maintaining one of the world's greatest upland wetlands. o

PARAMO DE SUNTARBAN, COLOMBIA
THE SECRETS OF ANDEAN GRASSLANDS

AT AROUND 3,000 METRES UP, the cobblestones were slippery from the near-permanent fog. It made progress slow on the ancient track climbing through the Colombian Andes. We were entering an improbable ecological and hydrological El Dorado known as the Páramo de Suntarban. *Páramos* are treeless, sunless and water-soaked tundra between the forests below and the glaciers above. Their saturated peaty soils were covered in weirdly shaped cactus-like plants, most of which are found nowhere else on Earth. The moisture infuses everything, seeping into lakes, peat bogs, springs, aquifers and ultimately into the rivers that supply Colombia's water. Without this and other páramos, the capital Bogota, and many other Andean cities, would be dry.

A few people work up here. On our journey up the track from the pretty white-walled 'last village' of Cácota, we passed fields as steep as staircases, where poncho-wearing farmers somehow managed to cultivate potatoes. There were some cattle, whose milk was left for collection in churns on the roadside far below. But humanity had been left behind by the time the path reached its destination. Visible fleetingly through the swirling fog was the Laguna de Cácota, a lake where all the moisture that the supersaturated soils and plants could not absorb had accumulated. Beyond, in the depths of the páramo, spectacled bears, pumas, Andean foxes known locally as 'páramo wolves', and tiny, endangered northern *pudu* deer, all reputedly lived. Above the clouds somewhere, Andean condors circled.

The páramos wrap around the cordilleras of the Andes. You find them between 2,800 metres up, where the trees stop growing, and around 4,000 metres, where the glaciers begin. Theirs is a unique waterlogged world with no seasons and minimal direct sunlight. Botanists treasure what are reckoned to be the most species-rich tropical mountain ecosystems in the world.[19] There are more than 3,400 species of plants in the Colombian páramos alone, including herbs, ferns, and more than fifty species of *frailejones*, spiky-leaved shrubs which can grow up to 15 metres tall and are covered in tiny hairs to catch moisture from the air.

Each páramo on each mountainside has its own species. About 60 per cent of them are endemic. Most plants are xerophytes, adapted to desert conditions because they can retain water. Water is the one thing these landscapes are not short of, but seemingly the plants love all they can get. Bathed in mists and spattered with rain, they are constantly sopping wet, and their roots sit in saturated soils that often hold several times their own weight in water.[20,21]

The richness of species in these alpine wetlands is an important reason why Colombia is one of the five most biodiverse nations on Earth. The plants are also of direct importance to millions of people downstream. They capture a huge amount of moisture from the clouds that constantly envelop them. The effective precipitation is, as a result, about 40 per cent more than actually falls as rain, said our guide, Sergio Ivan Niño, from the local conservation authority, CORPONOR. 'The whole place is a natural reservoir.'

The páramos provide 85 per cent of Colombia's drinking water from 2 per cent of its land area.[22] The capital Bogota, the biggest city in the Andes, gets most of its water from the Chingaza and Sumapaz páramos. The latter is the largest páramo in the world, covering around 2,500 square kilometres.[23] Páramos also water Medellin, the country's second city, Cúcuta on the border with Venezuela, and the tourist resort of Cartagena on the shores of the Caribbean. Dams on rivers draining páramos also provide Colombia with 60 per cent of its electricity.

But despite the huge importance of these alpine wetlands, we know remarkably little about them. They are the least-researched ecosystem in the tropics, says Wouter Buytaert of Imperial College London, who has studied them more than most. Some 90 per cent of the 70,000 square kilometres of páramos are in Colombia, but there are others in small isolated patches on mountains in Ecuador, Peru, Venezuela, Bolivia, and Costa Rica. Except for a handful of small outliers in East Africa and New Guinea, they are the only tropical alpine wetlands on Earth.[24]

Their wetness arises because the mountains of the Andes are buffeted both from the west by storms coming in off the Pacific, and from the east by hot humid air rising up out of the Amazon jungle. Both the Amazon and Orinoco rivers – which are respectively by volume the world's largest and third largest rivers – have their largest tributaries beginning in páramos. Take away the páramos and the Amazon rainforest would be much more prone to drought.

Nobody knows for sure what impact climate change will have on these critical wetlands. Maybe all will be well. Páramos have moved up and down the Andean mountainsides in the past, responding to natural changes in temperature. But the Colombian government says 75 per cent could disappear during this century.[25] Rainfall is already becoming more seasonal, with reduced humidity and cloudiness in the Colombian Andes, says Daniel Ruiz, of Cornell University in the United States.[26] The fear is that the carbon-storing, water-retaining, biodiversity-rich páramos could be replaced by a barren land devoid of vegetation and subject to drying soils and erosion.

There are other threats. Farmers are moving up the mountainsides, planting potatoes, herding cattle and draining the waterlogged soils.[27] 'If you destroy the soil, you destroy the water storage capacity,' said former Colombian Environment Minister Manuel Becerra. And there is a lot of gold and silver buried beneath the Colombian páramos. Those metals have remained largely unexploited over the past half century, because the FARC guerrilla group and drug barons took over most of these remote regions. Now that FARC has laid down its arms and the drug barons have lost much of their power, that is set to change.

The Colombian Government is keen to extract a peace dividend by bringing in mining companies. Ministers have promised that there will be no mining in the páramos themselves. But their own researchers at the Humboldt Institute told us ministers are deliberately excluding

FOLLOWING PAGES:
Thousands of species of plants occupy the páramo wetlands of Andean Colombia. Many, like these frailejones, with their long, spiky leaves, both thrive on the sodden soil and help keep it wet by capturing moisture from passing clouds.

remote páramos that should be protected. The institute's Carlos Sarmiento said those areas include the world's seventh largest undeveloped gold reserve. Much of the 600-square-kilometre La Colosa open-pit mine project is on a páramo. 'It's a direct conflict: water versus gold.'

Back down the páramo trail in Cácota, the village's Mayor held a small ceremony to thank his visitors. Villagers distributed peaches from their orchards and spoke in praise of their páramo. They hope to persuade the government to keep the mining companies out, they said. They have plans to develop eco-tourism instead. The walk up the ancient track could soon become a popular excursion, with the mysterious fog-shrouded lake as the prime destination. 'We can only have prosperity if we protect the environment,' said the Mayor. 'Our natural riches are our village's strength.'

SO FAR, REMOTENESS AND LAWLESSNESS have protected the páramos and the Laguna de Cácota. But many upland lakes and wetlands have been sacrificed to a pervasive belief that they are a waste of space. In Latin America, this view goes back right to the start of European colonization. It prevailed despite the experience of many pre-Columbian societies who managed such wetlands to deliver food to sustain advanced civilizations.

High up in central Mexico, 2,400 metres above sea level, Lake Texcoco was once at the heart of the Aztec Empire that dominated Central America from the fourteenth to the sixteenth century. Tenochtitlan, the imperial capital, was built on an island in the lake. The city was at the time probably the largest city in the world, with more than a quarter of a million inhabitants. Conquistador Hernán Cortés described it as another Venice. Its buildings were raised on stilts bored into the mud. Surrounding the city were tens of thousands of 'tillers [who] dwelt in the middle of swamps'.[28] Those tillers had turned the lake into a patchwork of fields surrounded by waterways. Cortés called these human-made wetlands *chinampas*, or floating gardens.

The *chinampas* covered around 90 square kilometres and fed most of the city's inhabitants. For a while, they allowed the Aztecs to withstand the Spanish invaders, who besieged the city in 1519. But while the Aztecs saw the lake and its floating gardens as a source of natural plenty, Cortés saw it as a barrier to his control over the Aztecs, and as a flood risk. So, as soon as he had vanquished the Aztecs, he set about draining the lake.

He and his imperial successors largely succeeded. Today, most of the waterways of the Aztec capital have been replaced by the streets of

Mexico City, one of the world's largest megacities with a population of more than twenty million. But some of the *chinampas* persist in the southern suburb of Xochimilco, at the far end of a new Metro line. Covering 22 square kilometres, they are less than a quarter of their former size, but they continue to produce fruit, vegetables and flowers for the city's bustling markets.

The *chinampas* are not – and never were – literally floating gardens. They are islands. Each field comprises layers of mud poured into the midst of the lake by the tillers, known today as *chinamperos*.

The *chinampas* 'floating gardens' in Mexico fed what in the fifteenth-century was the world's largest city, the Aztec capital of Tenochtitlan. The Spanish conquistador Hernán Cortés said it was like Venice – before sacking the city and draining the gardens. Today, just a handful remain.

Their banks are anchored by water-loving native willow trees. The *chinamperos*, some of whom still speak the old Aztec language of Nahuati, continue to keep the fields fertile by paddling down the 400 kilometres of waterways in narrow, flat-bottomed boats before each planting season, scooping mud from the lake bottom onto the gardens in buckets attached to long poles.

Some of the gardens grow five crops a year.[29] These days, the cabbages, celery, beetroot, purslane and green tomatoes are supplemented by beds of gardenias, hibiscuses and roses, sniffed by the city-dwellers who have turned the gardens into a picnic spot in a crowded city. Many of the *chinamperos* supplement their gardening income by paddling the visitors down the canals.

Most foreign tourists in Mexico City visit the stupendous Aztec pyramids and wonder at the power of an empire that built them as centres of worship and for human sacrifice. But at least as remarkable are the wetland gardens that fed that empire. They may be among the most fertile and longest continually cultivated areas in the Americas.

ELSEWHERE IN THE HIGH PLAINS of the Americas, other pre-Columbian societies transformed seasonally flooded grasslands into vast expanses of croplands, by creating what can best be described as corrugated land. Surrounding Lake Titicaca on Peru's southern border with Bolivia, 3,800 metres above sea level, today's altiplano pastures are criss-crossed by long narrow strips of slightly elevated dry land, interspersed by wet furrows. Though long abandoned, these former raised fields were known to the Quechua people as *waru waru*.

Their true purpose was divined by Clark Erickson, an archaeologist from the University of Pennsylvania, in the 1980s. He pointed out their similarity to the *chinampas*, and concluded that they were created by pre-Inca societies, who piled soil onto linear mounds, leaving ditches that functioned as irrigation channels. The *waru waru* around Lake Titicaca covered 820 square kilometres.[30] First tilled 3,000 years ago, they may well have been where the first potatoes were cultivated.

Many researchers were initially sceptical of Erickson's interpretation of the purpose of these strange corrugated landscapes created out of wetlands. To prove his point, he employed local Quechua farmers to construct fields on the lake's north-west shore, using traditional Andean tools. He then employed them to farm the fields. Yields were typically three times those in surrounding fields. Some years his fields produced crops when those on conventional fields failed because of floods. The system proved so successful that aid groups, archaeologists, and others have raised money to promote the system more widely.

Erickson and others have since found similar large expanses of corrugated land in the Bolivian Amazon and the Guianas, often close to substantial human settlements, on islands in flooded landscapes. In places, the furrows appear to have been dammed with brushwood fences that trapped fish living in the seasonally flooded grasslands. All evidence of elaborate well-organized pre-Columbian agricultural societies prospering in wetlands with a level of sophistication rarely achieved today.[31] ○

LAKE LOKTAK, INDIA
HOW HYDROPOWER BROKE THE MIRROR OF MANIPUR

F OR ANYONE FLYING INTO IMPHAL, a fast-growing city in the mountains of north-east India, Lake Loktak dominates the valley below. Shimmering in the sun, it is full to the brim with crystal-clear water crossed by fishers in dugout canoes and dotted with the lake's most unique features – floating islands. Hundreds of them. Unlike the *chinampas* of Mexico, these islands genuinely float most of the time, cruising across the water according to the winds. They vary in size from a few metres across to hundreds of metres. Some are more than a hundred years old, with accumulated spongy, peaty soil 2 metres deep, most of it hiding below the water line. These mini-wetlands have been found to contain at least 128 plant species,[32] though the dominant vegetation on most is beds of reeds as much as 5 metres high. Hidden among the reeds on the largest of the islands are the last of a unique sub-species of deer, known as *sangai*, whose hooves are adapted to traversing their spongy surfaces. Once thought extinct, there may be as many as 300 of them.

This magical vista is the physical heart and cultural soul of Manipur, a small mountainous Indian state, of which Imphal is the capital. Locals call the lake the 'mirror of Manipur'. With the once-rebellious state now opening up to foreign tourists, Lake Loktak is a jewel waiting to be discovered. But there is another side to this story, one not included in the tourist brochures. The lake's beauty is superficial and its ecology, including the floating islands, is in deep trouble. The problem is a large concrete barrage built just downstream. It blocks the river that once channelled the lake's waters into the River Irrawaddy, across the nearby border with Myanmar. Effectively, the barrage turns the natural lake into a human-made reservoir. The question now is whether it is too late to restore the lake to its former natural glory.

Until the Indian Government completed the Ithai Barrage in 1983, the size of Lake Loktak fluctuated dramatically through the year. It swelled when monsoon rains poured out of the surrounding mountains. But in the dry season, the water level in the lake fell and its surface area contracted by four-fifths. The floating islands settled for a few months onto the lake floor, where their roots absorbed nutrients.

Only the centre of the lake remained flooded. The rest was reduced to a series of pools surrounded by wide wetlands where fish bred. The lake, often covered in lotuses and water lilies, was a rich source of fish, harvested using techniques often unique to Loktak. The wetlands also grew wild rice and an array of herbs and spices from pennywort to shell ginger, while offering rich pastures, grazed by water buffalo. The entire lake ecosystem directly sustained the livelihoods of some 30,000 fishers and their families, and was the cornerstone of the state's economy.[33]

The barrage has changed all that. It keeps the largest lake in north-east India full all year round, so that it can deliver a constant stream of water through turbines to generate hydroelectricity. The natural flood cycle is no more. The barrage has permanently submerged the two-thirds of the wetlands that once emptied seasonally, and has prevented fish from migrating into and out of the lake. The 'mirror of Manipur' is today artificial, its apparent naturalness a mirage, said Ngangom Sanajaoba of the Loktak Development Authority, as we began a tour around the lake. His authority was theoretically in charge of maintaining the lake's health. But it had no control over what was making the lake sick – the barrage.

The *pengba*, an endemic carp, which once bred in the lake and which Sanajaoba calls 'the pride of Manipur', may now be extinct in the wild. Other famous local fish such as the *sareng* catfish and *khabak*

The shimmering waters of Lake Loktak in north-east India have long sustained the local Meitei population. This girl is searching for lotus flowers to sell. But the lake's ecology has been in decline since its outlet was dammed for hydroelectricity.

FOLLOWING PAGES: Thousands of fishers once spent much of the year camped on Lake Loktak, harvesting its produce using indigenous techniques. But the lake's ecological balance has been lost, and the government has banned living on the lake.

have gone too, he said. Wild stocks of other fish are a fraction of what they once were.³⁴ Numerous water plants once harvested by locals have disappeared too, including a wild sugar cane called *singmut*. The proportion of their food that the people of Manipur get from the lake and its wetlands has fallen from 60 per cent to 10 per cent, Sanajaoba said. Meanwhile, the lake's famous floating islands, known locally as *phumdis*, are slowly dying. The constant high water means they are permanently afloat. Unable to absorb nutrients from the lake bed, their soils are becoming ever thinner – often now too thin for their

LEFT: Meitei fishers on Lake Loktak persist with their traditional fishing methods, some of which the government has attempted to ban.

OPPOSITE: Organized fish farming is spreading across Lake Loktak in place of traditional methods of harvesting the lake's native fish species.

most illustrious inhabitants, the *sangai*, to walk on them safely.

As we journeyed down the western shore of the lake, it was clear that the reliably high waters were also allowing a human invasion. Flooded former pastures around the lake are now occupied by fish farms, which encroach ever further into the lake. They cover an estimated 100 square kilometres. A new road was being built when we visited, past a once-remote island temple and out into the lake. Venturing across, we saw excavators removing *phumdis* to create more open water that entrepreneurs would turn into fish farms. One plot along the new road was advertised for sale as a 'tourist development'.

The largest expanse of phumdis covers 40 square kilometres in the south of the lake. In 1977, it was declared a national park – the world's only floating national park – to protect the *sangai* that live there, sharing the reeds with wild boar, civets, and rare Indian pythons. But, starved of nutrients by the constant high levels of water in the lake, even these large phumdis are deteriorating, said Sanajaoba, as we surveyed them from a viewing platform. Only a fifth are now strong enough to support the *sangai*.

Words such as *sangai* and phumdi that describe the lake's unique features all come from the language of the local indigenous group. The Meitei make up around half the population of Manipur. 'Our culture was built round the plants, the fish and the wildlife of the lake,' said Sanajaoba, himself a Meitei. 'The waters used to have nineteen varieties of floating rice, which we harvested. The lake was once often

covered in flowers, and there were lots of bees. It was great for honey collecting. But all that has gone now.'

Some cultural traditions persist. In markets we visited around the lake, trading was still carried out almost exclusively by old Meitei women. They include Khwairamband market in Imphal, reputedly the largest market in the world run entirely by women. The markets are now mostly reduced to selling tiny smoked fish, and water chestnuts, a local delicacy. We bought a bag on a small quay, before taking a boat onto the lake, where we drank tea in a café erected on a phumdi. We watched a constant stream of canoes, some loaded with fishing nets and others ferrying people to a small community on Karang Island, the only solid-rock island left in the lake that is not reachable by road.

After tea, we visited a family living in a hut on a small phumdi. They use the hut, made of bamboo and plastic sheeting, as a base for fishing the lake the woman of the hut, Phajatombi (not her real name), told us. They have another home on the shore, but it is an hour away by canoe, she said, so they often stay on the floating island for several days. They had a mattress in the hut, and a small stove where she was smoking fish they had caught the previous night. There were a few carp – escapees from the fish farms – but most were minnows. A kitten played outside on the spongy vegetation, waiting for any scraps.

Thousands of fishers once lived on the phumdis. Their huts were made of the reeds that grew around them. Besides catching the fish that followed the phumdis, they grew vegetables in the phumdi soils. It was a complete, largely self-contained life in an enclosed aquatic world where the fishers were the undisputed masters. But for the past thirty years, the government has been in conflict with the Meitei, diminishing their livelihoods and ways of living.

The antagonism began with Meitei opposition to incorporating their former kingdom into India, which only formally happened in 1972. It was aggravated by the construction of the barrage, with its destructive effects on their way of life. Political opposition grew to militancy. Meitei gangs holed up on the floating islands before attacking government targets. The police and military hit back with brutal force. At one point they used hovercraft to spread out across the lake and clamber onto the phumdis. In flushing out insurgents, they killed many innocent people, said local human rights lawyer Babloo Loitongbam. He is coordinating a campaign at the Supreme Court in Delhi to call the state forces to account for more than 1,500 allegedly extra-judicial killings, many of them carried out on the lake.[35,36]

The conflict took on an overtly environmental aspect with the passage in 2006 of the Loktak Protection Act. It declared living on phumdis to be illegal, because constructing the reed huts damaged the mini-ecosystems. The charge is dubious and the constitutionality of the law is contested, Loitongbam told us. The Supreme Court in Delhi is considering whether floating islands should qualify as communal land, with protected rights of abode. But undeterred by the legal uncertainty, the government in 2011 removed some 700 families from the floating islands, and burned their homes to prevent them returning. Thus the 'mirror of Manipur' became a hostage to a conflict in which environmental managers found themselves accomplices in destroying local livelihoods.

Since the mass expulsions, people such as Phajatombi and her family have slowly gone back to living part-time on the floating islands, while catching fish. Occasional expulsions still happen but, by 2018, an estimated 300 huts were back on the phumdis.

The returning fishers have persisted with traditional Meitei fishing methods, some of which the government has also outlawed. These

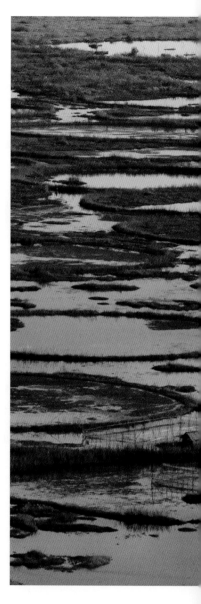

Floating masses of vegetation, known as phumdis, grow naturally on Lake Loktak. But locals have adapted them to create tiny reserves where they raise native carp species.

include *athaphum* fishing, a method that involves fishers weaving plant material gathered from phumdis into circular enclosures up to 200 metres in diameter, within which they capture and nurture carp. The enclosures are the best way of catching the lake's depleted fish stocks. At one point, there were more than 8,000 of them across the lake. There are fewer now, but their distinctive circular shapes are highly visible when flying over the lake into Imphal.

Today, there is an uneasy truce over the lake. But the sad truth, Loitongbam told us, is that 'a lake that used to be managed by the local people has been taken over by technocrats from outside'. The Meitei's traditional methods of exploiting the wetlands are no longer respected, or even acknowledged. 'The Loktak Protection Act contains no provisions for protecting the people who know it best. Their interests are not taken into account. And outside management has not been good for either the lake's people or its ecology.'

This takeover of the lake was made possible by the new hydrology of the lake, created by the Ithai Barrage. It ended the hydrological cycles that the Meitei knew best how to exploit, and created a lake ripe for a different kind of economic development. Arguably one that is much inferior.

To end our tour, we drove south beyond the lake to a gorge where the barrage, a mass of concrete and steel 60 metres across, has for nearly forty years blocked the main natural exit of water from the lake. The water on the lake side of the barrage was around 3 metres higher than downstream. Sanajaoba pointed to a hill above the barrage. It was known to the Meitei as the home of the Goddess of Wealth, he said. But that was not what the barrage had now delivered to his people. Environmental economists had calculated that for every one rupee of wealth generated by the hydroelectricity from the lake, five rupees had been lost due to the decline of the lake and its wetland ecosystems. The trade-off was a disaster for the lake and its inhabitants.

Some local people want to remove the barrage and bring back the old lake ecosystem. But Sanajaoba said that whatever the mistakes of the past, such a move would cause a new round of conflict, and new victims. 'After the barrage was built, people adjusted their livelihoods towards fish farms.' The fish farms now produce more than 20,000 tonnes of fish a year, many times the current wild fishery in the lake. 'If we changed back, all the fish ponds would dry up.'

But there could be a compromise. Since 2008, the Loktak Development Authority, with advice from Ritesh Kumar, Director of Wetlands International South Asia, has been trying to negotiate an agreement with the hydro-engineers that would open the barrage for a short while each year, to allow an annual managed drawdown of the lake. Even a few weeks of lower water levels would dramatically revive the lake's wetlands and wild fisheries, according to Kumar. The phumdis could be reconnected with the lake floor, and fish could migrate into the lake. He argues that the electricity grid in north-east India is now much less reliant on constant supplies of hydropower from the lake, so a suspension of supplies for a few weeks would not be disruptive.

It is an intriguing prospect. The Loktak Development Authority has proposed a range of options, including opening the barrage to allow the lake level to drop by just over 2 metres during May, sufficient to allow 42 per cent of the phumdis to be briefly grounded on the lake bed.[37] ○

RIFT VALLEY, ETHIOPIA
EAST AFRICAN LAKES DRAINED FOR A VALENTINE ROSE

Abule Debele is proud of his fields. They cover only half a hectare – just a couple of small patches of cabbages and onions. But they are his life's work. 'This is my asset,' he told us. 'I got it from my father. I won't sell it to anyone.' When we visited, the cabbages were almost ready to be cut and trucked to markets in Addis Ababa, the Ethiopian capital 130 kilometres to the north. But Debele, a courteous middle-aged man who had put on his best shirt for our visit, was agitated. He and the other farmers of the fields around us were tinkering urgently with a broken pump. The pump drags water up from the nearby Lake Ziway, which at 30 kilometres long is one of the biggest lakes in Ethiopia's Rift Valley and a vital water source. Without the water, the fields here are useless. And on the morning of our visit, the pump was refusing to work.

Debele and his fellow farmers are all members of a local farming cooperative that runs the intake pump and a network of underground pipes that feed the water into furrows along each farmer's fields. They have to maintain a strict rota system that has been in place since Debele's father's time. Debele gets water to irrigate his two fields for eight hours every four days. But the pumps are being worked ever harder, because the lake's shore has been retreating, and every year they have to extend their pipes further. 'This land was part of the lake once,' Debele said. 'Now the lake is in the far distance.'

We looked towards the horizon. The water was more than 2 kilometres away. It would retreat further during the dry season.

'I am worried the lake will get so far away that we can't get water to our fields, even when the pump is working,' said Debele. He does have a well in his field, as a back-up, but the underground water is increasingly salty. In any case, he said, 'as the lake goes down, so does the underground water level.' He pulled back the weeds around the well and we peered in. 'There used to be water 5 metres down. Now it is 15 metres, and we sometimes have to dig to 30 metres,' he said.

There was a sound in the distance as the pump resumed its work. Debele looked relieved. For now, he had water. 'I get two crops a year. If the lake was higher, I could get a third crop in the dry season,' he said. That is his dream, but he was not confident that would ever happen. It was more likely that one day the lake would be gone and there would be no water at all. For that reason, he doesn't want any of his six children to take over his fields. 'I want them to have proper jobs, so I have sent them to college,' he said.

There are thousands of farmers like Debele in the Central Rift Valley of Ethiopia. Altogether, they cultivate around 75 square kilometres of the valley, and get through some 150 million cubic metres of water a year, most of it from Lake Ziway.[38] They are major

OPPOSITE: Lake Abijata is in rapid retreat as water is abstracted from the interconnected lakes of Ethiopia's Central Rift Valley. Demand comes from both the world's largest rose farm, owned by a Dutch company, and evaporation ponds run by a soda ash manufacturer.

RIGHT: One of more than 5,000 irrigation pumps operated by local farmers around Lake Ziway, the only freshwater lake in Ethiopia's Central Rift Valley. Because of over-abstraction by smallholders and horticulturalists, the lake is retreating, threatening both their supplies and the lake's ecosystem.

water users and understand that the lake is getting smaller each year. Debele's cooperative is exploring more efficient irrigation systems that would drip water from pipes direct to the roots of the plants, rather than flooding their fields. That might reduce their demands on the lake – or just allow farmers to grow a third crop.

It would be easy to blame poor water-guzzling farmers such as Debele for emptying the lake. But that would rather miss the role of another farmer on the shore of Lake Ziway. Whereas Debele makes a meagre living from half a hectare of land and a share in a Yamaha pump, a short way around the lake, closely packed together, sit several

hundred giant greenhouses. All are humming with the sound of dozens of industrial pumps sucking up the lake's water. The complex covers 7 square kilometres of the lake shore and is owned by a Dutch flower company, Sher. It employs thousands of workers, drawn from all over Ethiopia. They pick a staggering four million rose stems every day, all of which are trucked out and flown to a depot near Amsterdam, from where they are sold across Europe. That is more than a billion roses a year. These greenhouses are Europe's premier rose garden. Many of the roses carry the Fairtrade logo.[39] But you have to ask: fair to whom? Arguably not to Debele and his fellows, who some would say have a prior claim to the water.

LAKE ZIWAY IS NOT JUST a handy water reservoir. It is also home to Ethiopia's second largest fishery, and an important wintering ground for water birds migrating from Europe. Flamingos were there in profusion during our September visit, attracting tourists staying in lakeside hotels. But the lake is in trouble. Its average volume has declined by 200 million cubic metres in recent years, thanks mostly to the constant suction of pumps around its shore.[38] Hayal Desta of Addis Ababa University suggests that it could one day follow the fate of Lake Haramaya 300 kilometres to the north-east, which has dried out entirely within the past twenty years, literally drained to death.[40]

To lose the lake would be devastating for almost all areas of the local economy, said Nugusa Ushe, Head of Environment at the local district council, who accompanied us on our tour. Already, fish catches have fallen by two-thirds, he said. Those that are netted often 'look diseased and discoloured' because of pollution that he believes comes mainly from fertilizers and pesticides used in Sher's greenhouses. 'We advise people not to eat them,' he said, as we stood on the lake shore opposite the greenhouses. An impromptu gathering of local fishers agreed. 'We catch the fish from the lake that they pollute,' said one, pointing at the greenhouses. 'There are fewer fish today. It damages our livelihoods.'

The greenhouses are far from the only reason for the lake's problems. Debele and his fellows play an important role too. But Sher is by some way the biggest single abstractor from the lake. We wanted to know exactly how much water the company's greenhouses take from the lake. Sher does not declare a figure and did not respond to our questions. But Herco Jansen of Alterra, a Dutch hydrological consultancy, calculated its annual water use in 2007 at 20,000 cubic metres per hectare, which, over 7 square kilometres, would add up to 14 million cubic metres. That is equivalent to about 10 litres per rose, and remains the best publicly available estimate.[38]

The Ethiopian Government, which launched Sher's business on the lake by selling the company an old state farm, clearly thinks that the water makes an important contribution to the country's economy. But others question that. At the end of the day, lives do not depend on growing flowers in the way that they do on growing food. If hard choices have to be made, an income for Debele to feed his family ought to have priority over a Valentine's Day gift to a sweetheart in Europe.

Four million roses are cut every day in the greenhouses of the Sher Ethiopië company, the largest abstractor of water from Lake Ziway.

Hard choices may have to be made. Lake Ziway is the most upstream of three interconnected lakes in the Central Rift Valley, a closed basin with no outlet to the sea. It is the only one of the three that contains fresh water. It has a single outlet down the River Bulbula. As water levels in Lake Ziway have fallen, the flow down that river has declined by around 75 per cent.[38] And that is bad news for another ecologically important water body that relies on regular inputs from Ziway: Lake Abijata.

Lake Abijata is a soda lake famous for its thousands of pink-plumed lesser flamingos.[41] Bird fanciers come from across the world to see the birds, which breed more than 500 kilometres away in Kenya's Lake Nakuru, but come to Lake Abijata to feed on the blue-green algae in the lake. The Deputy Director of the Ethiopian Wildlife Conservation Authority, Solomon Mekonnen, told us, 'Lake Abijata is unique. You can see 200 species of birds in a day there. And with peace and security in the region increasing, everyone is coming. Foreigners, but people from Addis too.'

But Lake Abijata, too, is in retreat. In 2015, its surface area shrank below 100 square kilometres, less than half what it was in the mid-1980s. In the three years before our visit in 2018, the shore receded a further 400 metres. The decline is not explained by any change in rainfall. It is partly a response to reduced flows down the River Bulbula from Lake Ziway, and partly due to its own giant water abstractors. In particular, since the early 1980s, huge pumps have been taking its water to evaporation ponds that extract soda ash, sodium carbonate, which is widely used as a water softener. The plant, half owned by the Ethiopian state, has twenty-two evaporation ponds that stretch for 6 kilometres around the lake's shore. They evaporate 150,000 litres of water to make every tonne of soda ash.[42]

When production was at its peak, the ponds consumed 13 million cubic metres of water from the lake each year.[43] That couldn't last. By 2006, the lake shoreline had retreated beyond the reach of the main intake pumps. Activities have continued more fitfully since. On our visit, the lake was a kilometre away and only five ponds had water in them. The only staff there were guards who said they were on permanent watch to prevent sabotage by locals, who blamed the operation for the state of the lake and the lack of fish. Plans to rehabilitate and expand operations with more intake pumps have, thankfully, not come to anything.

As Lake Abijata has shrunk, the salt content of its water has more than doubled, poisoning the algae on which the flamingos feed. Flamingo numbers have declined. Even if the soda ash plant remains

PREVIOUS PAGES:
Pink-plumed lesser flamingos fly in formation over Lake Abijata where they feed on the soda lake's blue-green algae. Ornithologists come from all over the world to see them.

largely out of operation, it still faces the loss of water from Lake Ziway. 'Abijata could be gone in a decade,' said Wetlands International's lead for the Rift Valley, Amdemichael Mulugeta. He is busy trying to devise solutions based around pragmatic plans for sustainable water management in the basin.

Can the lakes of the Rift Valley be saved? The answer should lie with the Rift Valley Lakes Basin Authority. Its Director-General, Kebede Kanchula, sounded up for the task. He was so keen to see us that he opened up his office in the lakeside town of Hawassa on a Saturday morning. 'The lakes are only 2 per cent of the country's land area, but they produce water for 100 million people,' he said. Lake Ziway, as 'the only freshwater lake we have', is key. Its main economic activities – floriculture, agriculture, food processing and tourism – all demand water, which is scarce everywhere, he said.

Despite the grand title of his authority, Kanchula said that he has no powers to rein in water use. And that includes the actions of some government agencies. The Ministry of Agriculture is still trying to expand irrigated agriculture, he said. He hopes that devolution of power being carried out under the new Prime Minister Abiy Ahmed can give him the powers he needs. Perhaps before the lakes start drying out completely.

He may find unlikely allies in small-scale farmers. Debele's cooperative is part of a wider union of some 150 local cooperatives, which operates throughout the Central Rift Valley. The Meki Batu Fruit and Vegetable Growers Cooperative Union represents 9,000 farmers tilling some 40 square kilometres, most of it irrigated.[44] The union provides seeds, fertilizer, machinery, and animal drugs, as well as getting farmers credit lines, training, maintenance of pumps, and providing centralized marketing of their products. But, its manager Yadecha Abera told us everything its members do depends on water.

The union has decided that it cannot wait for the government to act to protect its members' most vital raw material. Abera told us it was introducing its farmers to drip irrigation. It hopes that widespread adoption of the simple water-saving technology will reduce water needs for farmers by 25–30 per cent. It could save a million cubic metres of water a year – provided the farmers don't respond to the water savings by growing more crops. And provided, of course, that Sher and other bigger water users take similar steps.

It looks like a way forward in a region where rising water demand is coming up against absolute water limits. Mulugeta at Wetlands International spoke of the organization's commitment to partner with the Meki Batu Co-op, the Basin Authority, Sher, and others.

The aim is to proof the landscape against future hydrological disaster by improving the water efficiency of smallholder farming, by helping bring these groups together to share knowledge, and jointly drawing up a water allocation plan that would maintain the Bulbula River, ultimately ending the decline of the lakes. It should serve as a model for other lake systems all down Africa's Rift Valley.

Mulugeta was at pains not to blame anyone for the current water crisis in the valley. But at the Basin Authority, Kanchula did single out the flower industry as in particular need of a hydrological makeover. When Sher had arrived in the country, it had been given carte blanche

58 WATER LANDS

The Gibe III dam in Ethiopia is Africa's tallest. As it diverts water for sugar plantations in the Omo Valley, it will largely dry up Lake Turkana, the largest desert lake in the world, on which half a million people over the border in Kenya depend.

by a previous government to grow its business, he said. But times are changing in the Rift Valley. Water security for smallholders now matters at least as much as selling roses to Europeans. 'Sher needs to become water efficient,' he said. 'It needs more advanced technology. I have been to greenhouses on Lake Naivasha in Kenya where they are using hydroponics to save water. They could do that here. Sher says it is costly, but it's not beyond them with a business like that.' With luck, a new way forward for water management in the Central Rift Valley is emerging.

GREENHOUSES AND OTHER water-guzzling enterprises are a growing threat to a number of ecologically important water bodies and wetlands all along East Africa's Rift Valley. They include Lake Natron in Tanzania, where plans for soda ash extraction threaten its flamingo colonies; Lake Naivasha in Kenya; and perhaps most dramatically Lake Turkana, 300 kilometres south of Lake Ziway, just over the border in Kenya.

Lake Turkana is the largest desert lake in the world, and stretches for 250 kilometres through remote, arid northern Kenya. Half a million people depend on its waters, including the Turkana people after whom it is named. But the lake could be largely gone within twenty years, because the river that supplies 80 per cent of its water from over the border in Ethiopia was dammed in early 2015. The 243-metre-high Gibe III dam, the tallest in Africa, blocks the River Omo. Its primary purpose is to generate hydroelectricity, but it will also allow water to be diverted to irrigate sugar and other thirsty crops on new farms being constructed in the Ethiopian bush.

The project is designed to transform one of the most remote regions of Ethiopia. According to people in the distant capital of Addis Ababa, the Daasanach and Mursi tribes of the Omo Valley are backward. They scorn tribal women who wear traditional giant plates in their lips.[45] The government wants to use the power of water to change them. That means generating hydroelectricity, and converting some 3,000 square kilometres of their floodplain pastures into irrigated commercial farms. But if things go to plan, little of the river's flow will be left for Lake Turkana.

The government is in a hurry. The reservoir behind Gibe III began filling in 2015. The 1,000-square-kilometre Kuraz sugar plantation was already under construction a year later. Another 500 square kilometres is earmarked for a cotton farm. But the downsides are already visible on the river's floodplain. Farmers say that the dam has ended the natural flow downstream of water and silt during the wet season. As a result, woodlands and pastures across the floodplain are drying out, soils are deteriorating, and crops have been poor.[45] Village life is suffering. Traditional tribal festivals such as the bull-jumping initiation ceremony for boys have been cancelled. And things are even worse over the border.

In the first three years after the dam was closed, water levels in Lake Turkana fell by almost 2 metres. That was enough to dry out the shallow Ferguson's Gulf on the lake's western shore, where most of its fish feed and grow. Until recently, the Turkana people harvested up to 18,000 tonnes of tilapia annually from the bay, selling smoked fish at beach markets and for export as far as the Congo. But as the lake empties and the Gulf's shoreline retreats, fish catches have already declined sharply.[46]

This is just the start. Sean Avery, a hydrologist at the University of Leicester, estimates that the diversions of water to the planned Ethiopian farms will soon cut the supply of water and silt to the lake by up to 50 per cent.

Within twenty years the level of the lake is likely to drop by more than 15 metres, says Avery. That would be sufficient to reduce it to two small, salty pools. The three national parks in and around the lake, where hippos wallow and Nile crocodiles breed, would lose most of their wildlife.

The hydro-plant is a done deal. While it could be operated to provide a short seasonal flood pulse, there is little chance of it being shut. But, says Avery, the farms that will divert water out of the river altogether are still works in progress and could be halted. That could yet save Lake Turkana, the world's largest desert lake, along with the livelihoods of its fishers and the pastoralists and traditional farmers along the river. o

MOSCOW PEATLANDS, RUSSIA
FROM PEAT MINES TO BOG WILDERNESS

Zoya Drozdova was delighted to have a Western environmentalist in her car. Russians don't often get to show the world a thing or two about nature protection. But she knew she had a winner. For she was about to show off one of the world's largest restored peat bogs. Some 150 square kilometres of moorland in the Meschera National Park east of Moscow that had been deforested, drained, and partially scraped of its peat by an industrial juggernaut of the former Soviet Union, has now been returned, under her watch, to something like its former glory.

'We are restoring nature, and making a place where people want to come,' she said, as we drove down a track that for many decades had been a railway bringing thousands of workers into the bog, and taking millions of tonnes of peat to feed the nearby Shatura power station. The area all around us had been a black industrial landscape. But in 1992 the peat workings of Meschera were shut and incorporated into a new national park. There, Drozdova and her colleagues set to work restoring the natural bogs.

They made up the methodology as they went along, she remembered. At first, they had constructed large concrete dams on the main drainage channels to raise water levels. But the dams created such big differences in water levels within the bog that the peat they were trying to save was instead being eroded. So, they switched to a gentler method of rewetting. They blocked small drainage channels instead, often simply by bulldozing peat into them. It worked. As of late 2018, they had installed thirty-five small dams and almost all of the bog had been restored. There are pine and birch trees on the higher ground, and oak, willow, and black alder in the boggier places. Park staff have taken to calling the black alders, with their roots bathed in water, 'Russian mangroves', said Drozdova.

The Meschera park is today a jewel of biodiversity. Drozdova reeled off the list as we got out of the car for a stroll. 'We have 900 species of plants, 160 of lichens, seventy of mosses, including twenty-three sphagnum species, fifty species of mammal and 215 of birds, more than a hundred of which come for the water.' We jumped over creeks that for many years had drained the peat but are now full of fish. Along the paths, we saw several grass snakes. A large beaver lodge forms the focal point of a nature trail. An area the size of metropolitan Brussels has been turned back to nature. At less than fifty dollars per hectare, Drozdova said, it had not even been expensive.

RUSSIA HAS AROUND A QUARTER of the world's peat bogs. They cover 1.4 million square kilometres of the country, an area larger than France, Germany, and the UK combined.[47] The largest expanses are in the frozen tundra of Siberia, but the river floodplains around Moscow are richly covered in peat, too. Though long valued and harvested by locals, outsiders often regarded the bogs as useless wastelands. Until the early nineteenth century, when Tsar Alexander I promoted the idea of draining them to create farmland for Russia's growing population. Then, new railway companies began to cut peat to provide fuel for their steam locomotives. And in the years after the 1917 Bolshevik Revolution, Vladimir Lenin saw local peat as the ideal fuel to generate electricity in far-flung communes across the country.[48]

Lenin recruited fellow Bolshevik Ivan Radchenko, who had worked in an early peat-fired power station, to mastermind the programme. At first, the peat was cut by teams of women working in near-slave conditions, knee deep in freezing water and wielding little more than kitchen knives. They were brutal times. Radchenko eventually fell out

Russia's peat bogs are being restored. Following peat mining, this wetland in the Tver region of Russia, near Moscow, was for decades a flat, black wasteland.

with Lenin's successor, Joseph Stalin. He died a prisoner in a gulag in 1942. But his programme to electrify Russia with peat continued. By the 1940s, the women peat-cutters had been laid off. Instead, mechanical excavators dug drains into the bogs to dry out the peat, so that it could be milled by heavy machinery. By the 1980s, production had reached 150 million tonnes a year. Around 90 per cent of all the world's peat extraction was being done in the Soviet Union.[49]

After the collapse of Communism, the industry nosedived, leaving behind tracts of drained peat covering an area almost the size of the Republic of Ireland. Scrub vegetation began to grow on the exposed dry peat. During droughts, they became tinderboxes. In the summer of 2010, during a record heat wave, wildfires on former peat workings almost encircled Moscow. They burned for months, even smouldering through the winter beneath a thick layer of snow. The burning peat created black smog that shrouded the Russian capital and killed thousands of people.[50,51]

The authorities hurriedly brought in outside help. They found it particularly from the German Chancellor, Angela Merkel, who was concerned that the peat fires were releasing huge amounts of carbon dioxide into the atmosphere, accelerating global warming. She sent experts who recommended rewetting the bogs to stop the fires and safeguard the carbon. At a ceremony in 2011 in Hannover, Merkel and then-President Dmitry Medvedev launched the Russian Peatlands Restoration Project, known as PeatRus. It was partly funded by the German Government's International Climate Initiative,[52] and covered four former peat-extracting provinces around Moscow: Tver, Vladimir, Novgorod, and Moscow itself. It aimed initially to rewet 400 square kilometres.[53] Wetlands International, a partner in the PeatRus project, reckons it could eventually be extended to 1,000 square kilometres.

At the start, money seemed to be no object. Engineers in Moscow province built large concrete dams and elaborate sluice gates to control the bogs' waters. In Meschera National Park, Drozdova and her colleagues scoffed. The lesson they had learned a decade before, that low-tech ecological methods of rewetting work best, was ignored. 'After 2010, a huge amount of money was wasted. It was money burned. We spent much less money and didn't have their problems,' Drozdova remembered, as we concluded our tour of her bog paradise.

The lesson is now being learned afresh. There is no need for complex hydrological engineering. A bog restored according to ecological principles does not require engineers to constantly measure water levels and open and close sluice gates. It runs itself. 'It takes some time to restore the ecological balance, but it is the most sustainable and cost-effective process in the long run,' says Frank Mörschel of the German Government's development bank, KfW, which has managed finance for the PeatRus project.[54] He has good news on the climate front, too. He estimates that every hectare of rewetted bog reduces annual carbon dioxide emissions to the air by between 5 and 10 tonnes. At fifty dollars per hectare, that is one of the cheapest ways of curbing carbon dioxide emissions devised anywhere.

THE DEVELOPMENT OF CHEAP, effective, sustainable, and ecologically desirable methods of rewetting peat bogs and staunching their carbon

A smog crisis in Moscow in 2010 was mostly caused by fires in abandoned peat workings in the countryside around the city. The lethal pollution triggered a programme to rewet the peat and prevent future fires.

dioxide emissions is great news. There are abandoned peat bogs requiring restoration all across the flat lands around Moscow. And the restoration movement is taking off.

We visited the Orshinsky Bog, which covers 670 square kilometres of floodplain on a tributary of the River Volga in Tver. Between the 1920s and 1980s, peat extractors drained a third of the bog. After the works closed, 'they used to have fires in the cut areas all the time', said Vladimir Panov of the Tver State Technical University, who has been in charge of ecological rehabilitation. But after five years, they have

raised water levels by up to a metre across 65 square kilometres. Work continues. We watched a bulldozer move peat to block a drainage channel about 2 metres wide, close to a maze of newly formed lakes. 'It was dry here till a year ago,' Panov said. 'You can see the soil is getting waterlogged again. The reeds are new; sphagnum is spreading too.'

Not everyone is happy, however. Villagers around Orshinsky Bog have always gathered mushrooms and berries in the bog, and fished its lakes. That continued in the undrained areas even when peat extraction was at its height. Now, the recent rewetting has submerged some of the roads that the villagers used to reach the lakes. 'They don't like us,' said Panov. 'They have attacked us and last year they damaged a tractor.' The conflict seems a shame. Surely conservationists and local bog harvesters have the same ultimate aim, to preserve its biological wealth? Panov agrees. But he does not welcome people he regards as intruders, and has no plans to reprieve their roads. 'They need to understand that what we are doing here is good. Anyway, fishing here is illegal,' he said.

But the battle of Orshinsky Bog is an exception. More often, we found that bog restorers wanted the locals to appreciate and use their works. Officials of Kameshkovo, a town in Vladimir, were keen to show off their restoration of abandoned peat workings. Project designer Victor Bilanko had just finishing blocking the drains with dams of peat and wood when we visited. In two months, water levels in the peat had risen by 30 centimetres, he told us. The bog vegetation was returning; so were fish. He designed his scheme so that townsfolk could easily get to the lakes to fish. The resurrected bog was already a valued part of the community.

Near the Dubna River, a tributary of the Volga in Moscow province, we visited an area where much of the peat had been drained for agriculture. But not all. Amid the fields one wetland survived. The Taldom Homeland Nature Reserve has been the haunt of Russian ornithologists since the 1920s, said Olga Grinchenko, the reserve's current Director. The ornithologists had influential friends and it was saved from being drained. Now, the tide has turned. Many of the old collective farms are being abandoned, and conservationists are reclaiming the fields for boggy nature. Grinchenko and her colleagues have created nature trails. 'People love to come. We have elks and bears here, and rare orchids,' she said.

SO, WHAT SHOULD BE DONE with these recreated bogs? Should they be rigorously protected inside national parks? Or opened up for harvesting by locals? Or both? Russia is not densely populated, but even so conservationists cannot assume they will get their own way. We went back to the Meschera National Park, which has been rewetted the longest, to find out what lessons they had learned.

Drozdova and her colleagues had inherited a peatland that was far from empty of people. 'There were forty-six settlements inside the park boundaries, made up of people who had mostly been employed in the peat workings, but who had a variety of other activities,' she said. 'So, we held meetings in every village. Farmers were worried about their fields being flooded; hunters wanted to know if there would still be ducks to shoot. We reassured them that they would be able to

High-school student Alya returns down a muddy track to her home in the half-deserted village of Tasin in Meschera National Park. 'Nobody has jobs here now,' she says.

continue. In the end they backed us with the rewetting, because they all wanted the fires to stop.'

The truth, however, is that without work extracting peat, many people don't want to stay. There are some jobs in tourism, and servicing the park. But it is remote. A road through the park that heads for Moscow is often impassable because of rising water levels. People are voting with their feet. In twenty years, the population has halved to around 10,000.

One village we visited, Yagodin, has just two permanent residents. Most houses are occupied for only a few weeks each year, by holidaying Muscovites. Another, Tasin, wasn't much better. It once had 3,500 inhabitants, in homes kept warm by a peat-fuelled heating plant. But now the heating plant is derelict, its pipes struggle uselessly across the village, and just 200 people live in a few dozen dilapidated wooden houses. Many of the houses look empty. We met sixteen-year-old Alya, a high-school student returning home on her bike down a muddy track. 'Nobody has jobs here now,' she said. 'I have no father and my mother doesn't work. We have little money so we rely on the bog for many things. I go picking berries all summer. But when I finish high school I will have to go to Vladimir [a city 60 kilometres away] to find work.'

More than 97 per cent of Russia's peat extraction has ended in the past thirty years.[49] But in many places, there is talk of small operations restarting, usually to supply horticulture rather than power stations. Down a long, rutted road, deep in birch forest near the village of Sukhoverkovo, south-west of Tver, we found production had resumed. Vegetation had been removed across a clearing of a thousand hectares, and the old drains deepened. Excavators scraped dry peat into giant piles, where it awaited trucks to take it away.

The new Director of Operations, Ignatiev Valeriy, sees a bright future. Before the site was abandoned, the old Soviet operators had removed the peat to a depth of 2–3 metres. 'But the peat has a maximum depth of more than 5 metres, so there is plenty left for us,' he said. 'We have enough for fifteen years.' In the first two years, they had extracted 150,000 tonnes.

Was this a return to the bad old days? Valeriy defended his operations. People need jobs, he said. And anyway, 'we are producing material for making Moscow green. It will be a growing medium for plants.' Tatiana Minayeva of Wetlands International gave him guarded support as we tramped over the site. 'They chose this place because it was an abandoned old peat working. We say that if Russia is going to carry on extracting peat, it should be done in places like this, rather than at new sites.'

As the sun set over the birch forests around us, a convoy of big excavators returned to their compound, the drivers took off their hard hats and mounted their motorbikes. They headed past the big Alsatian dog guarding the gate, down the lanes through the forest to their villages. Some would probably stop off to go fishing in the ponds that still dot the area. They were happy: there was work again on the old peatland. ○

SIBERIAN MIRES
PERMAFROST THAWING AND CARBON UNDER THREAT

PEAT FORMS WHERE DEAD VEGETATION fails to rot, usually because the ground is waterlogged. In Siberia many areas have been accumulating peat for 10,000 years. Peatland in such remote and inhospitable places is rarely drained for farmland or fuel. But there are other threats, such as the invasion of humans to find oil and gas.

Western Siberia – a region stretching for 600,000 square kilometres from the Urals to the Yenisei River and from the shores of the Arctic Ocean to the Russian steppes – contains one of the world's largest peatlands.[55] And it has some of the world's largest and most ill-managed oil and gas fields. Tens of thousands of drilling rigs and their attendant infrastructure are scattered across huge areas. The combination is not pretty. A watery landscape unwalked since the days of the mammoths by anything heavier than a reindeer, is now broken into fragments by pipelines, roads, flood embankments, pylons, seismic survey lines, and other paraphernalia of hydrocarbon extraction.

The oil industry has traditionally regarded the Siberian mire ecosystems as being without value. As it commandeered the terrain, it for a long time made no attempt even to collect the different roads, pipelines and survey lines into corridors to minimize their intrusion. Instead, they each made their independent way, with land often cleared for 200 metres or more on either side. Even road networks were established by different organizations for their own purposes. Helicopter journeys frequently reveal three or more roads and tracks heading in the same direction within a few hundred metres of each other, each road perched on top of its own embankment, blocking rivers, dividing lakes, disrupting drainage and thwarting animal migrations.

Some years ago, during a visit, officials in the oil town of Noyabrsk said the surrounding area had 12,000 oil wells, each discharging into open lagoons around 40 million cubic metres of polluted water per year.[56] As much as a tenth of the discharges ended up forming acrid films that glistened on lakes and mires. And there was air pollution. Clouds of black smoke fell to Earth from indiscriminate flaring. The reindeer were long gone, and even most of the birdlife had departed. Since then, there have been overtures to instigate change.

Wetlands International has developed guidelines for minimizing such impacts in Arctic wetlands, which it has also inventoried. Even so, the mires are often littered with an inheritance from years of mismanagement, abandoned high tension cables, rusting pipes and broken machinery, while streams of oil and chemical pollution from oil wells and leaky pipelines seep down backwaters and pond up in the mires. Much of the oil heads down long pipelines that cross the Urals, headed for the task of heating homes in Western Europe. Yet the apocalyptic scenes that attend this supply have failed to attract condemnation because they are so remote that few people see them.

THE SECOND GREAT THREAT to the integrity of the Siberian permafrost is global warming. A while after our visit to Noyabrsk, an email arrived from a Siberian ecologist with what he called 'an urgent message for the world'. Sergei Kirpotin of Tomsk State University had made an expedition across thousands of kilometres of West Siberia. It was a region he knew well, but he had discovered a rapid change in just the previous three years. Huge areas of permafrost were suddenly thawing. Spongy areas of lichen and moss were turning

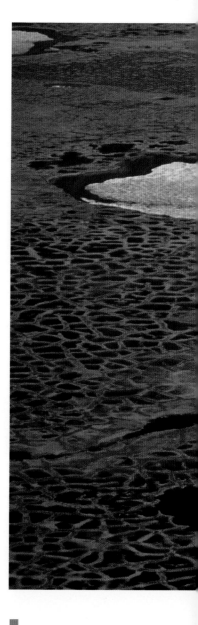

Naturally formed mires in the Siberian peatlands. Changing land use, including the spread of oil and gas production, risks exacerbating thawing caused by climate change and releasing billions of tonnes of greenhouse gases into the atmosphere.

into a landscape of lakes. 'We have never seen anything like it, and had not expected it,' he wrote.[57] He called it 'an ecological landslide that is probably irreversible and is undoubtedly connected to climate warming'. Thus: His concern was that, in combination with land use change, the thawing permafrost would start releasing its stored carbon in the form of the greenhouse gases carbon dioxide and methane.[58]

The Cold War satirical movie *Dr Strangelove* imagined that Russia had built a nuclear 'doomsday machine' in the heart of Siberia. It was the ultimate fictionalized version of a real-life military strategy known

as mutually assured destruction. If one nuclear bomb was dropped on the Soviet Union, the doomsday machine would automatically unleash a thousand more, killing all humanity. The film ended with its anti-hero, Major Kong, plunging from an American bomber towards the tundra, astride a device that would obliterate the planet. Climate pessimists fear thawing permafrost in Western Siberia, the world's largest store of peat carbon, could be almost as deadly. Kirpotin was warning that he believed he was seeing the first signs of the apocalypse unfolding.

By some estimates, a quarter of all the carbon stored in the world's wetlands, soils and forests is in the peatlands of Western Siberia.[59] Alarmingly, the largest pools of peat are in the southern areas most susceptible to thawing. Canada, Alaska and northern Scandinavia also have big frozen reserves. The total store of carbon accumulated in permafrost peat is estimated at around 1,700 billion tonnes,[60] twice the amount in the atmosphere.[61] If global warming continues to thaw the permafrost, this carbon could bubble up – as carbon dioxide if oxygen is present, or as methane if it is not.

Recent studies suggest that the release is already under way. Kirpotin's 2005 discovery has proved to be just the start. Surface thaws are resulting in the rapid multiplication of lakes on the surface of the permafrost. The lakes are fast-tracking methane to the atmosphere, according to Katey Walter Anthony of the University of Alaska Fairbanks.[62] During the summer, 'large plumes of bubbles' release the gas from the lakes into the air. The lakes can prove temporary. In some regions they drain away as the permafrost thaws to greater depths. But while they persist, she says, 'this hotspot bubbling is unlike anything that has been observed before.'

Meanwhile there is growing concern that the processes will be accelerated by land-use change, the physical collapse of permafrost soil as it thaws,[63] and by wildfires of the kind that broke out across the region during the exceptionally warm summer of 2019, covering an area larger than Denmark.[64] The permafrost carbon may be released into the atmosphere over decades rather than centuries.

Climate modellers suggest 2° C of summer warming in the tundra regions could add almost 50 per cent to summer emissions, creating a potentially dangerous positive feedback loop as the emissions add to warming, which in turn releases more emissions. One assessment reckons such feedback could bring us to the UN limit of 1.5° C of warming (above pre-industrial levels) about 20 per cent more quickly than the world's burning of fossil fuels alone.[65]

PREVIOUS PAGES:
Rigs extracting oil litter the peatlands of Western Siberia, causing widespread pollution as well as fragmenting wetland ecosystems and disrupting the local hydrology. This image was taken near Sugut in 1993.

FROM THE SOUTHERN TIP of Argentina to the vast expanses of Siberia, from Ireland to Indonesia, and from Poland to Burundi, known peatlands cover an area of the planet larger than India. Wherever dead plant material accumulates in waterlogged soils, there is peat. Peatlands make up around half the world's wetlands. In some places it is home to reindeer and moose, and in others to tigers and orangutans. In Greece's Philippi Bog, it is 190 metres deep. The stricken Chernobyl Nuclear Power Station in Ukraine is built on the Pripyat peatland, which is now rather radioactive. More than a quarter of Finland is underlain by the stuff. Most of the 1982 Falklands War between Britain and Argentina was fought on peatlands that, according to the International Union for the Conservation of Nature, 'may represent some of the least disturbed soils on the planet', containing more carbon than in all of Ireland's bogs.[66]

Peatlands were long regarded as somehow 'nature's mistake', says Katja Bruisch, an environmental historian at Trinity College Dublin. They were 'associated with backwardness, in economic and cultural terms, as well as with a lack of reason and order', fit only to be drained and put to some useful purpose, whether farming or peat extraction.[67] Now we know differently. Besides their ecological value, peat bogs and mires are the planet's biggest natural stores of carbon. A hectare of peat typically contains around 650 tonnes of carbon for every metre of depth, around twice as much as any forest growing on it. In some countries, they are the largest stores of carbon, or second only to fossil fuel reserves such as coal and oil. In Britain, the peat bogs of the Flow Country in northern Scotland – a bleak, treeless and drizzly landscape often compared to Arctic tundra – store an estimated 400 million tonnes of carbon,[68] equal to four years of the country's fossil fuel emissions. The region is probably the largest area of blanket bog in the northern hemisphere, and one of the most intact. Environmental economists put the value of 'Britain's rainforest', through climate regulation and other services, at £220 million a year.[69]

While peat remains waterlogged, it generally stays intact. But when drained, as it dries out oxygen reacts with carbon, releasing it into the atmosphere in the form of carbon dioxide. Globally, drained peat may be emitting 2 billion tonnes of carbon dioxide a year, as much as industrial emissions from the whole of India. And fires, such as those in Russia in 2010, dramatically speed up this process of oxidation. For twenty-five countries, emissions from drained and burned peat are equal to more than half the national emissions from fossil fuel burning.⁷⁰ Those countries include Indonesia, Mongolia, Myanmar, several East African countries, and peaty European countries such as Iceland, Finland, and Sweden.

One sensible strategy for protecting drained peat from fires and oxidation is to plug up the drains and restore the wetland. The good news is that many countries are doing just that. Russia's rewetting is far from a one-off. Wetlands International is a founder member of the Global Peatlands Initiative, aimed at restoring peatlands across the tropics, and is arranging carbon finance, under the Paris climate agreement, to restore peatlands in Indonesia that have been damaged by the spread of palm oil and other agriculture.⁷¹

On the Baltic shore of north-east Germany, almost 3,000 square kilometres of peatland in Mecklenburg-Vorpommern have in the past been drained for agriculture, including intensive cattle grazing. But a decline in cattle numbers in the region following the reunification of Germany in 1990, combined with rising concern about the cost of draining marginal agricultural lands, led to a rethink. A tenth of the lost peatland has since been rewetted, preventing estimated carbon dioxide emissions of some 300,000 tonnes per year.⁷² Reeds growing in the restored peatlands are now being burned as a green energy source in power stations.

Will the same happen in Ireland? The country has long been known for its peat bogs, which once covered almost a fifth of the country. While used as a fuel for a thousand years or more, the peat has since the eighteenth century been a vital household fuel, cut by hand and dried in the sun. In the 1930s, the process was mechanized in the same way as in Russia. By the 1960s, peat burning provided 40 per cent of Ireland's electricity – a much greater proportion than was ever achieved in Russia. Peat areas such as the Móin Alúine, in the centre of the country between the rivers Shannon and Liffey, were criss-crossed by narrow-gauge railways to remove the peat.

But that is all coming to an end. In 2018, the state peat harvesting company, Bord na Móna, announced that all seventeen of its remaining bogs under active peat extraction would close within a decade.⁷³ The move is intended to cut the country's carbon emissions. But without a programme to rewet the bogs, the drained peat will continue to emit carbon dioxide, perhaps for centuries.⁷⁴ The company is committed to rewetting 100 square kilometres of former workings in the coming decade.⁷⁵ Continued at that rate, shutting down emissions from dried-out former peat workings could take many decades. Pressure is bound to grow to accelerate the rewetting

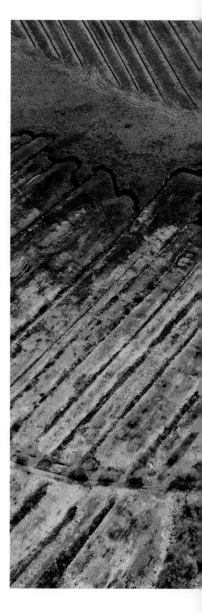

Seen from above, the landscape of the Bog of Allen in Ireland is severely scarred by industrial peat cutting.

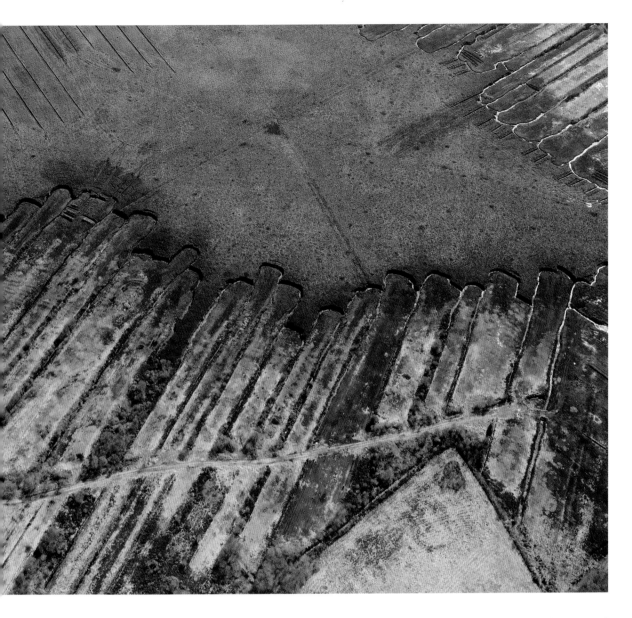

programme. While full restoration is seldom possible, rehabilitation of the country's peatlands, where possible, will help to lessen the carbon they emit.

In Europe, with the support of the Dutch Government, Wetlands International and the International Peat Society have engaged with the peat industry to help establish the Responsibly Produced Peat Foundation, whose certification system protects areas of high conservation value, while securing the best possible use of former peat extraction areas, for recovery of ecosystem values.[76] ○

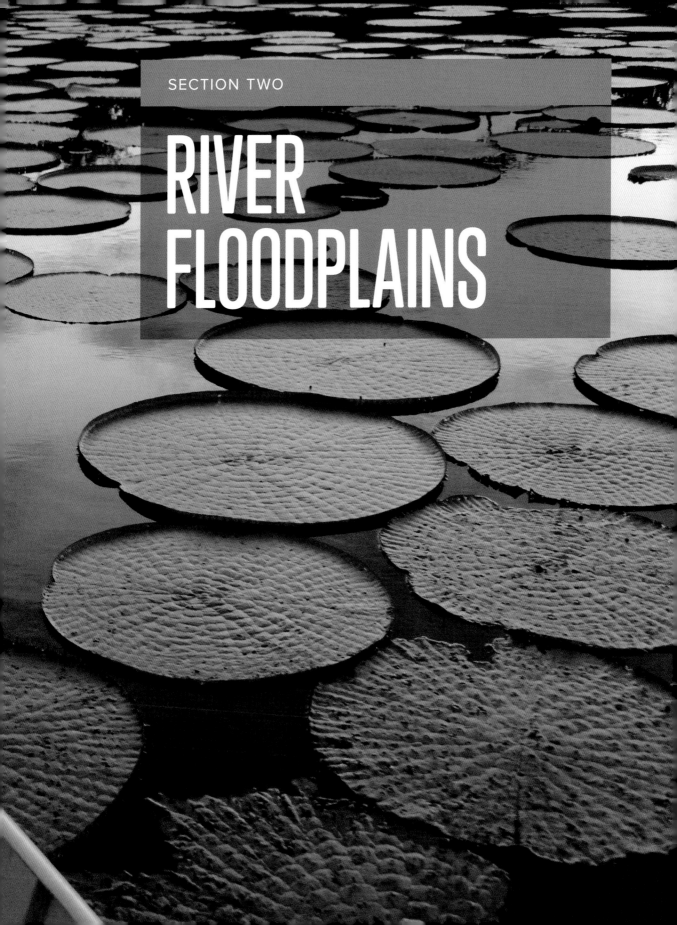

SECTION TWO

RIVER FLOODPLAINS

Rivers flowing out of uplands eventually find flatter ground. If left untamed, they meander and spread across wide floodplains that are among the world's richest ecosystems. 'The floodplain is like a body, with the natural channels and seasonal rivers its veins, the water its blood; if the flow is blocked, it gets sick,' says Aboukar Mahamat, a Cameroonian academic who grew up in a fishing family near Lake Chad.[77] Many are sick, thanks to dams, drains and dykes. But from the Pantanal of Brazil and the Tonle Sap in Cambodia, to the Inner Niger Delta of Mali and the Mesopotamian Marshes, local wetlanders live on – and hold the key to their revival.

PREVIOUS PAGES: A boat on the Rupununi River, which drains the seasonally flooded lands of the Wapichan people in southern Guyana. The territory is a hydrological meeting place, where creeks that flow south into the Amazon mingle with others that flow north to the Atlantic.

LOCATIONS FEATURED IN SECTION TWO

ENGLISH FENS AND THE AMERICAN MIDWEST
LAND GRABS IN 'DISMAL SWAMPS'

Take a train across the flat plains north of the English university town of Cambridge and you see an almost entirely human-made landscape. Where once there were wide expanses of reeds and willow trees, meandering rivers, backwaters full of eels, and skies festooned with geese and storks, now there is a symmetry of rectangular fields and canalized rivers running straight as a die towards the North Sea. Only from the air can you catch an occasional glimpse of the past, remembered in ghostly curves of whitish silt amid the black peat. Curves that mark where the old rivers once ran through an area still known as the Fens.

It was the Romans who first tried to remake the land here. Their military engineers drained some areas and dug canals across the quagmires to speed grain-carrying boats to garrisons in northern England. After those works silted up, natives spent the Middle Ages harvesting the Fens, rather than draining them. The pickings were rich. Eels, caught by the hundred in wicker traps, were a speciality. The eleventh-century Domesday Book records a local economy so dependent on eels that the great Fenland cathedral city of Ely got its name as the 'eel district'. Taxes, rents, and tithes to the churches were all paid in eels. Meanwhile, chroniclers reported that salmon and sturgeon could be found in such quantities 'as to cause astonishment in strangers'.[78] Fleets of coaches took the fish and fowl to London.[79]

The Fens covered an area of some 3,900 square kilometres, and the Fenlanders lived in a world apart. To outsiders, their wet empire was an alien wasteland, plagued by mosquitoes[80] and governed by complicated customs. The Fens were common lands, and the eels, fish, and wet pastures were commonly owned resources. By the early seventeenth century – an age of enclosure and widespread privatization of common lands across England – this had become an affront to the landed gentry. Among them, the Earl of Bedford wanted the Fens to grow grain for sale in London. To do that, he had to break the Fenlanders' grip on this dank, mysterious, and dangerous place. So, in the 1630s, he and thirteen other 'adventurers' formed the Bedford Level Corporation and concocted a scheme to drain the Fens and divide the land between them.

The engineer they chose for the job was Cornelius Vermuyden, a Dutchman who had successfully promoted himself in England as the *éminence grise* of Dutch drainage engineers using newfangled windmills for pumping. Though why, if that was the case, he'd had to leave Holland was never explained. In any event, since arriving in England in 1621, Vermuyden had caused havoc across the country with botched drainage projects. When he rebuilt flood defences on the Thames, he left them 'in a worse condition than before', according to an inquiry.[79] He had spent a brief stretch in prison after failing to pay for repairs to another project at Hatfield Chase on the River Trent, which had also caused more flooding than it had prevented.

Undeterred, his plans for the Fens were as ambitious as anything ever attempted in England or Holland. He proposed digging six cuts – giant ditches the size of rivers that would receive water scooped from the sodden ground by thousands of buckets strung from wheels powered by windmills. The meandering River Ouse was to be replaced by the largest of the cuts, named the Bedford River after the Earl. The eels, fish, and reeds would be replaced by cattle and grain.

The commoners fought back against an ill-disguised land grab that would destroy the waterlogged wealth on which they relied. They 'fell upon the Adventurers, broke their sluices, laid waste their lands, threw down their fences and forcibly retained possession of the land', as one local report put it.[81] This was guerrilla warfare in the Fens, fought with pitchforks and spades, and it went on for years. They had a song too, aimed at castigating the adventurers.[82] The following lines are two extracts.

> For they do mean all fens to drain
> And water overmaster,
> All will be dry and we must die,
> 'cause Essex calves want pasture
> ...
> We must give place (oh grievous case)
> To horned beasts and cattle,
> Except that we can all agree
> To drive them out by battle.

Despite the opposition, the infrastructure for the first 400 square kilometres of the scheme was completed, and Vermuyden was paid. But the windmills and sluices failed to do the job. And with natural drainage disrupted, flooding was often greater than ever. Both Fenlanders and their lords and masters were now up in arms against Vermuyden. So, the King stepped in. In 1638, Charles I reappointed Vermuyden to put things right, by supplementing the barely completed Bedford River with a New Bedford River. Charles named his terms for investing in the scheme: 200 square kilometres of drained land and a promise from Vermuyden to build him an 'eminent town' on a hill overlooking the two Bedford rivers. It was to be called Charlemont.[78] It would be a Venice of the Fens.[83]

The new project proved to be Charles's undoing, however. Among those who took up arms against it was a young man who had inherited land in the Fens but took the commoners' side. His name was Oliver Cromwell. His revolt against the King's drain turned into a revolt against the King's rule itself: the English Civil War. It went well. By 1649, Cromwell had vanquished Charles and chopped off his head. But Cromwell then changed sides. In 1653, three days after he had assumed dictatorial power over the country, he sent the army to the Fens to suppress his former friends.[79]

Cromwell finally finished the New Bedford River in 1652, using the labours of a team of Dutch prisoners captured in a naval battle the previous year.[78] But the wetlands got their own back on the traitor Cromwell. He died in 1658 from malaria, caught not in the Fens but while fighting the Irish in the bogs of their own land. Meanwhile, Vermuyden – rogue, charlatan, and sometime engineer – left the Fens

PAGE 82:
Fenlands in European landscapes, often shrouded in early-morning mist, are rare reminders of a bucolic past.

FOLLOWING PAGES:
Eel fishing on the Fens of eastern England. Peter Carter, pictured, says his ancestors have caught eels here for 500 years, using traditional wicker traps.

for Somerset in the West of England, where he sought financiers for a scheme to drain another great English wetland, the Somerset Levels. But there were no takers. As more and more of his embankments and drains across England failed, and aggrieved landowners demanded their money back, Vermuyden hid from public view. His final days seem to have been spent in poverty under the assumed – and somewhat ironic – name of Cornelius Fairmeadow.[84]

What of the Fens? Much of the drained farmland gradually shrank as its peaty soil dried. In time, it sank below the level of the cuts that were supposed to drain it. The windmills were not up to the job of keeping the fields dry. The entire enterprise was only saved by the arrival of the steam pump in the 1820s.[85]

Today, the two giant Bedford rivers still run, dead straight, across the Fens towards the sea. The fields are drained and deliver crops. But at what cost? Successive British governments have avoided revealing how much it costs to keep the pumps going and stop the Fenland fields from becoming waterlogged once again. But slowly the folly of the present state of affairs is dawning. The Great Fen Project, supported by government bodies such as Natural England and the Environment Agency, aims to rewet 37 square kilometres by blocking the drains.[86] A new wilderness of reed beds and eels, inhabited by birds such as the bearded tit, cuckoo, and bittern is returning. All it now needs is the return of the Fenlanders.

THERE ARE OFTEN GOOD REASONS for draining boggy ground. They range from eradicating malaria to protecting crops from waterlogging and providing solid foundations for homes. The practice goes back a long way. Even before the Romans, the Etruscans were draining Italy's coastal wetlands. Across Europe the Middle Ages saw piecemeal drainage from the Marais Poitevin on the west coast of France to Italy's Po Valley, before the Dutch began to create almost an entire nation from drained wetlands (which we return to in a later chapter).

Early European colonists in North America, who went in search of furs, found their richest pickings of beavers, minks, and muskrats among the continent's swamplands. But later settlers decried what Charles Dickens, in *Our Mutual Friend*,[87] described as 'dismal swamp', full of 'crawling, creeping, fluttering, and buzzing creatures … a hotbed of disease, an ugly sepulchre, a grave uncheered by any gleam of promise'. Encouraged by rising prices for exporting grain to Europe, they set about draining the swamps. Soon, almost every town across the Midwest had factories churning out cheap ceramic drains and

Chinese attempts to drain Lake Dian, south of Kunming in Yunnan, ended in disaster. The state farm failed, and today the lake, once known as the 'sparkling pearl', fills with blooms of algae.

steam-powered machines for digging ditches and pumping water. Eventually, all but 5 per cent of the once-boggy Midwest was drained.

Much the same happened in the Deep South, where land behind the new levees along the Mississippi was systematically drained. Around half of the Florida Everglades, a wetland that once covered 10,000 square kilometres, was drained.[88] In the west, California's Central Valley succumbed. Only 10 per cent of the state's wetlands remain today.

But things did not always work out. Draining the more northerly wetlands often faltered, and the agricultural returns were poor. Many

investors went bust after the financial crisis of 1929. But by then the Federal Swamp Land Acts had turned drainage into a national endeavour.[6] Where the swamps defeated smaller private drainage companies, the Army Corps of Engineers took over.[89] In the two centuries from 1780, thanks in large part to federal funding, the contiguous forty-eight states lost 500,000 of their 900,000 square kilometres of wetlands. President Trump once called the federal government in Washington a 'swamp' that needed draining. But for most of America's history, the federal government has been America's biggest drainer of swamps.

As the twentieth century progressed, many developing nations decided to drain their swamps too, often with Messianic zeal. During Mao Zedong's Cultural Revolution of the 1960s and 1970s, China engaged in an orgy of drain digging to create new agricultural land. Lakes were a special target for reclamation, notably in Hubei in central China, ending its centuries-old claim to be the province of a thousand lakes. China's second largest freshwater lake, Lake Dongting on the Yangtze floodplain in Hunan, was halved from its original 5,650 square kilometres.

Most emblematic of the Maoist fervour for taming nature's waters was the destruction of the Haigeng Wetland on the northern shore of Lake Dian, just south of Kunming in the mountains of Yunnan. Surrounded by Buddhist and Daoist temples, it features in many old paintings. But history did not protect it when, in the spring of 1970, Mao launched a 'great military campaign' to 'build dykes, drain water and create earth' to help feed the nation.[90] Some 300,000 peasants, intellectuals, Red Guards, and factory workers filled in nearly 25 square kilometres of marsh to create the Haigeng Agricultural Farm.

The result was disastrous. The new fields were barren and almost permanently waterlogged. So little grain was produced that the farm was abandoned in 1982. Judith Shapiro of Columbia University, who told the story in her book *Mao's War Against Nature,* quoted a former worker on the state farm saying the project 'was a bottomless abyss. Everything disappeared into endless mud.'

Today, without the neighbouring marshes to cleanse its waters, pollution builds up in Lake Dian, which was once known as the 'sparkling pearl'. With no wetland breeding and spawning grounds, fish catches have fallen by more than 90 per cent. Birdlife has disappeared. The local climate has changed too. Many blame the hotter summers and colder winters on the loss of the moderating influence of the wetland and the reduced lake. Making the best of a bad job, the Kunming city authorities have now turned the abandoned farmland into a park. It is advertised as a 'lakeside bathing beach with green willows, sky-piercing eucalyptuses, and blooming flowers'.[91]

Well, it's something. But it's not a patch on our next subject, the world's largest freshwater wetland. ○

PANTANAL, BRAZIL
A FUTURE FOR THE WORLD'S LARGEST WETLAND

OUR HEADLIGHTS ILLUMINATED SOMETHING big and four-legged ambling across the road. We swerved. Spotting the danger, the capybara, the world's largest rodent, showed a remarkable turn of speed to leap off the road and into a lake. It was lucky. A few minutes earlier, we had passed the carcase of a giant anteater, occupying almost an entire lane of the highway from Campo Grande to Corumbá in Brazil – the main route across the southern fringe of the Pantanal, the world's largest wetland.

Roadkill is serious here. Zoologist José Sabino of Anhanguera-Uniderp University in Campo Grande has been travelling the BR-262 for thirty years. In that time, he has catalogued 1,400 squashed animals. They belong to eighty-four species, many of them regarded as endangered, such as the giant anteater, jaguar, giant otter, and maned wolf. But such carnage is still a mere pinprick in a landscape with a density of wildlife unseen today anywhere outside Africa.

The Pantanal occupies some 200,000 square kilometres in the heart of South America, straddling the borders between Brazil, Bolivia, and Paraguay. Ten times the size of the Florida Everglades, it is a vast and remote land of muddy water flows, lakes and seasonally flooded grasslands, interspersed with forest islands and salt pans. In the wet season, more than 80 per cent of the Pantanal floods. Ranch buildings and a few raised tracks are often all that poke above the water. But in the dry season, the rivers that cross it falter. The floodwaters slowly evaporate in the sun, leaving behind just a few areas of permanent lakes.

OPPOSITE: The black plumage of this neotropic cormorant, a denizen of the Brazilian Pantanal, appears reddish brown in the early morning sun as it roosts in a tree. There are greater populations of birds here than almost anywhere on Earth.

RIGHT: The jabiru, a stork with a wingspan of almost 3 metres, is a constant presence in the Pantanal – Brazil's and the world's largest freshwater wetland. It can catch and eat baby caimans.

FOLLOWING PAGES: The Pantanal, a vast and remote land of muddy water flows, lakes, and seasonally flooded grasslands, interspersed with forest islands and salt pans, straddles the border between Brazil, Paraguay and Bolivia.

The Pantanal is not pristine. Much of it is grazed by cattle wandering watery ranches the size of small European countries. But more than 85 per cent of its vegetation is essentially natural.[92] Its hydrology, determined by the seasonal flood pulse, is also still largely intact, shaping the landscape and determining what people can do there. Nature has so far been equal to most human attempts to tame it.[93] Back in the 1970s and 1980s, armed gangs shot upwards of a million caimans a year, mainly for their skins. But the reptile barely missed a beat. There are currently an estimated ten million of them, more than four per square kilometre.[94] Hunting is now outlawed, except for indigenous subsistence consumption. Caimans are instead bred in giant farms on the Pantanal, and their meat sold in supermarkets.

After our close capybara encounter on the road, we stopped for the night near the fishing community of Passo do Lontra on the River Miranda. The following morning, we took a boat ride with Fabio

Roque of the Federal University of Mato Grosso do Sul in Campo Grande, who has a research camp close by. Within minutes, as the sun rose, we saw dozens of caimans snoozing among the water hyacinths, trees weighed down with birds looking for fish in the waters below, and happy families of capybaras gambolling on the banks. Five giant otters swam out to meet our boat, growling when we came close to take pictures.

The Pantanal is not a great centre of unique species. Instead, it is a melting pot where species from the Amazon rainforests and *cerrado*

grasslands to the north meet those from the pampas and *chaco* plains to the south. The birdlife is phenomenal. It didn't take long to spot the wetland's ornithological icon. The jabiru is a stork with a wingspan of almost 3 metres. It specializes in catching fish that play dead in the muddy beds of dry-season ponds. But it will also eat baby caimans.

'There are more birds per square kilometre here than anywhere on Earth,' said Roque. Birdwatchers can feast their eyes on endangered greater hyacinth macaws and American rheas, the continent's largest bird, which cannot fly but runs at speeds of up to 60 kilometres an hour. There are huge colonies of migrating egrets, spoonbills, herons, and a lapwing with a cry that can frighten horses – all predated by hawks, vultures and, at ground level, by hungry anacondas.

Despite the Pantanal's size and fecundity, there are threats to its ecosystems. One, mentioned more than any other by the people we met, is big-city fishing tourists who charter speedboats and snub their noses at fishing quotas and the local fishers with their quaint old traditions of preserving fish stocks. On our lazy early-morning river journey, bankside animals scurried for cover whenever speedboats shot past. At Passo do Lontra we met the owner of one eco-tourist

RIGHT: On the Rio Miranda in the Pantanal, a giant otter comes to growl at the photographer. The species is endangered.

OPPOSITE: The Pantanal is an important stronghold for jaguars, the most water-loving big cat.

lodge who told us he had a separate lodge for the tourist fishers. 'The two don't mix,' he said. 'The fishers are loud and play music; the eco-tourists are quieter.'

The outsider fishers also have a reputation for being a mainstay of the prostitution business in Corumbá, the crumbling old colonial port town on the banks of the River Paraguay where we were headed to next. At a riverside restaurant, Reginaldo, a local part-time fisher and former soldier, blamed fishers from São Paulo and elsewhere for the collapse of stocks in the river. 'When I came here twenty years ago, you could catch a metre-long fish with a simple rod right here in town,' he said. 'Now I have to go two hours in my boat to get fish half that size. The tourists go out in fast boats to the best spots, including the conservation areas of indigenous communities. They don't care and nobody stops them.'

FROM A FIELD OF COW-PATS outside Corumbá, we hired a Cessna plane to fly north across the Pantanal. It is hard to imagine anywhere flatter than the Pantanal. It slopes only a few centimetres in each kilometre. Its water takes three or four months to find its way across the Pantanal from the Brazilian cerrado highlands in the north to the River Paraguay, which continues south and eventually enters the South Atlantic beyond Buenos Aires in Argentina. This slow flow means that the Pantanal's wet and dry seasons are staggered. The dry season starts in May in the north, but not until August in the south, which was the month of our journey.

As we flew, the landscape beneath us was remarkably varied. At first, the rivers dominated, meandering and bifurcating crazily, creating backwaters and shifting channels. But further north it was hard to see where the rivers were. The water was mostly in small lakes, coloured blue, green, or muddy black. Then came the near-impenetrable heartland of the Pantanal. The Taquari River spreads out in a great fan of water, silt, and sand that covers almost 40 per cent of the wetland. The river has recently become clogged with sediment washed down from eroding farmland in the cerrado. So much so that it is no longer navigable. Much of the water simply washes in a sheet across the land.

Further on, in areas where the dry season was more advanced, the vegetation was in places yellow. The soils were cracked and there was smoke in the air from grass fires. With less water, the land looked more lived-in. There were straight boundaries between fenced pastures, as well as tracks crossing the grass and skirting scrubby woodland. The tracks often met at ranch buildings, some of which had an air strip. This is cowboy country, but many richer ranch owners are at their ranches part-time at best, flying in and out, with their main homes in Campo Grande or São Paulo. We began to wonder, who owns the Pantanal? And what do they want from it?

THE PEOPLE WHO KNOW the Pantanal best are those who use it, and who understand the traditions of its exploitation. The majority are poor *pantaneiro*s, the dwindling numbers of indigenous people, and the descendants of African slaves and European migrants, who live in the wetland. They work as fishers and cowboys, or increasingly in new industries such as mining and servicing tourists. After landing at Porto Cercado on the north side of the Pantanal, we set off to find some of them, heading upstream on the River Cuiabá, the wetland's biggest river.

Climbing the river bank in front of a group of wooden houses on stilts, we met Leandro Ribeiro and his wife, Antonia. They were taking

A traditional fisherman in a dugout canoe on the River Cuiaba in the Brazilian Pantanal. Restrictions have been imposed on such local fishers after rampant overharvesting by tourists equipped with speedboats and freezers.

a siesta. Leandro was born in 1945 and has lived in the small fishing community of around thirty people off and on ever since. 'This is our land, even though we have no papers to prove it,' he said. His story is typical of many male pantaneiros. As a young man, he had joined his father to work on a local ranch. He wistfully remembered the highlight of the year, when they drove the cattle down bush tracks for slaughter in São Paulo. 'We took 7,000 animals from the seven local ranches,' he said. 'It took sixty-eight days. There were plenty of jobs for cowboys then. But not now; the animals go by truck.'

Leandro always fished as well. 'I started when I was twelve years old. We often went out at night. You could hear the noise of the fish, there were so many.' The fish are not so numerous today, he said. The decline began with the arrival of fishing tourists in the 1970s and 1980s. 'They were allowed to fish as much as they could. It was uncontrolled. They filled their freezers. They had a great time and made money too, selling the fish to big guys from the cities.'

The bonanza couldn't last. Fish stocks faltered. In 2014, the state authorities of Mato Grosso, which has jurisdiction over the north of the Pantanal, banned everyone from catching high-value fish, whether they were traditional pantaneiros or tourists. Leandro now seems resigned to losing this livelihood. He has a pension now. And with electricity supplied to their community, the Ribeiros' home has a TV, washing machine and freezer; and we saw two old his-and-hers Nokia phones on the kitchen table. Life is fine.

But his wife, Antonia, is more combative. 'There is still a lot of illegal tourist fishing. We should protest but we don't,' she said. Before adding with a conspiratorial smile that she is a good fisher too. 'You need to go to the right spot in the river, at the right time. You need to know the river well, but the fish are there.'

These days, pantaneiros are only allowed to catch small fish, often as live bait for use by the tourists. And they are entangled in red tape. There are long waiting lists for permits to fish, and quota systems that require lots of paperwork. They feel like they are paying the price for the sins of others. Wetlands International's Brazilian boss, Rafaela Nicola, agrees. 'There is a lot of prejudice against traditional fishers,' she said as our boat continued upstream. 'Big landowners, tourist companies, and their client politicians push the agenda that the pantaneiros are to blame for fewer fish. But it isn't true. The aim seems to be to end artisanal fishing altogether. But we try to support them.' She says that, left to their own devices, the local fishers can be the best custodians of the wetland's fish stocks. 'We believe they are part of the solution rather than part of the problem.'

WE NEXT WENT TO São Pedro de Joselândia, a much bigger settlement of around a thousand people, 12 kilometres from the River Cuiabá. The cattle-rearing community has a school, several churches of different denominations, an air strip, a horse track that holds an annual race for pantaneiros, and a host of entrepreneurial activities. Most of those activities seemed to us to be run by one man, Joison Advincola. He was nicknamed 'pica-pao', meaning woodpecker, because of his constant activity. We sat down for lunch in his café and bought snacks from his

FOLLOWING PAGES: There are still cowboys in the Pantanal, overseeing the herds of livestock that graze the wetland. But many of their old tasks, such as driving cattle to distant markets, are now outmoded. The cattle go by truck. Many old farmhands have now moved to the cities.

small grocery store. He has rooms for passing Pantanal researchers, and a fishing lodge by the river for tourists. He manages a small manufacturing plant that makes cement rings used to construct wells. He runs an informal bank for collecting state pensions for villagers, rents out cars, buys and sells cattle, and has a butchery business.

Despite São Pedro de Joselândia's distance from the river, floodwaters lap at its houses during the wet season. We saw as many horses as cars, and more boats than either. Behind a dried-out pond containing three canoes, we met Joaquim. At 106 years old, he was probably the oldest pantaneiro. Proud of his physique, he had combed his hair carefully before we arrived and sat bare-chested on his veranda. He has thirty grandchildren from eleven sons and daughters. Most of his family now live in distant towns, but he was still enjoying life in the Pantanal. Especially during the wet season. Yes, water sometimes gets into his house. 'But when the land is flooded, you can travel by boat all the way to the river, if you want,' he said with delight.

The settlement was a strange mixture of the old and new. Across the street from Joaquim, we met one of the Pantanal's last traditional healers. Altemisia Dias Rodrigues holds regular consultations, like any family doctor. Advincola, the entrepreneur, sends his children to see her. She hands out teas and herbs from the flora of the Pantanal. A pink-flowered tree in her garden provides an antibiotic she applies to wounds, as well as infusions to combat worms and asthma. 'But now I encounter things I can't deal with,' she admitted. Modern afflictions such as depression and criminality are beyond her ministrations, she said. She blames them on the spread of agricultural chemicals and processed food. She suffers from depression herself. 'I took doctors' pills for that, so maybe I am not such a good healer of myself,' she smiled.

Other traditional uses of the natural resources of the Pantanal live on. Wild food plants are still popular, according to Geraldo Damasceno-Junior of the Federal University of Mato Grosso do Sul, who has studied them. Black rice grows in profusion in lakes. The palm-like caranday trees provide fibre for hats sold in every roadside store. Outside Poconé, a town on the north side of the wetland, we met Jacinto Maximo da Silva, who makes rocking chairs from wetland timber and reeds. The rockers are built from the wood of the rare lixeira tree, he said. 'And I have a special place on a river near here, where I collect the reeds. I only collect after the full moon. The elders told me when I was young that if you cut them at other times, they will rot. And it is true.'

Jacinto is a lone craftsman now: the last artisanal furniture-maker, he thought. But his products are popular, bought by tourists and big shots from São Paulo. Other Pantanal cultural and craft traditions are making a comeback, too. Many are popular with both tourists and the many former pantaneiros who have lost their jobs in the wetland and wash up in towns such as Poconé.

We stumbled on a craze for the old pantaneiro dances, for instance. Like the pantaneiros themselves, the dances are a mix of indigenous, slave and European traditions. They are typically performed by twenty-four masked male dancers, half of them dressed in women's clothing. They are often accompanied by brass bands and songs, many about the wildlife of the Pantanal.

Nobody is sure of the origins of the dances, but they are popular in ranch communities in the wet season, when there is little work to do, said biologist Cátia Nunes da Cunha of the Federal University of Mato Grosso in Cuiabá. Now enthusiasts are reviving the tradition. In local schools, teachers train students to dance. They have a ready audience. In five days in the area, we watched four different troupes perform – in the well-appointed Poconé cultural and sports centre, during an evening street market in the town, outside a church in São Pedro de Joselândia, and in front of the pool at a tourist hotel.

One troupe wore t-shirts declaring that they were 'rescuing the culture'. Well, sort of. The dances have been tamed somewhat. Gone are the feathers from Pantanal birds that once adorned the dancers. The male characters no longer carry knives to protect their partners. Sometimes girls get to dance, too. And their social purpose is more grittily urban. 'The majority of my dancers are from broken homes,' said the Director of one troupe, art teacher Natalice Soares da Silva, who had been brought up with the dances when she lived in the Pantanal with her gun-toting mother. 'If they didn't go dancing, many of them would be drug dealers.'

Many pantaneiros are descendants of the indigenous communities who exploited the resources of the wetland long before Europeans showed up. Those communities lost huge areas of their land in the mid-nineteenth century when ranchers first moved in on a large scale after wars in the region. There was genocide, says Nunes. While many indigenous people have since become integrated into the wider community, some still live in their old villages on the edge of the Pantanal, harvesting its resources and demanding their land back. The Brazilian Government has formally recognized indigenous territories currently covering around 3 per cent of the Brazilian Pantanal, but they claim far more.

Traditional Pantanal dances, such as this one on display in a school in Cuiabá, are being revived – often as a means of keeping urban youth out of trouble. They are a favourite with tourists and locals alike.

The Terena people, for instance, regard 460 square kilometres along the River Aquidauana as theirs. The territories contain their graveyards and are clearly marked by chalk signs long ago inserted into holes in the ground, said Terena teacher and activist Celma Francelino Fialho. Her people have gone to court to demand the return of their lands. They have won a number of cases where the chalk boundary marks were clear. 'Once we get our land back, we will reforest it and regenerate the water sources,' she said. But other groups may lack the cohesion to fulfil such aspirations.

The 1,500 Kadiwéu people, the survivors of a tribe that once occupied much of Paraguay and Brazil, still control more than 5,000 square kilometres of the southern fringes of the Pantanal. They too want to restore water sources and sacred areas – including a place with Pau Santo trees, whose resin makes a black pigment they use in their much-prized pottery. But their situation is complicated. Some families have leased their areas to outsider cattle ranchers. Internal disputes over the way ahead have led to killings. Wetlands International's Brazilian staff have been working with the tribe to help them develop management plans for their land.

THE PANTANAL IS COMPLICATED both ecologically and socially. So, what is the way forward? There are three main competing claims to the huge areas of watery land within the Pantanal. One, clearly, is that of the indigenous communities. A second is conservation. These two claims – the oldest and the most recent – have much in common, but sometimes clash.

Ecologically protected areas in the Pantanal are often made up of land from which traditional residents and resource users have been excluded, often under violent circumstances. In particular, a cluster of private reserves in the north-west of the Pantanal, created in recent decades, took over ranches 'on the understanding that people living within the areas would be removed before the land was transferred,' says Nicola. Anthropologist Rafael Morais Chiaravalloti of University College London agrees.[95] 'Local fishers claim that during the creation of the protected areas, they were displaced from their original settlements and ... restricted from using fishing sites of fundamental importance for their livelihood,' he reported.

Some of the reserves are resorts for high-rolling eco-tourists. Critics claim others are land grabs by mining companies and others. But there is another side to the reserves. The biggest private reserve, covering 1,080 square kilometres on the River Cuiabá south of Poconé, was assembled in 1997 through the purchase of seven large interconnected ranches. Its ecological work has allowed the reserve to become a Ramsar site. But it also has a social remit. The reserve is owned by Social Service for Commerce (SESC), an organization set up by some of Brazil's major employers, which also built, adjacent to the reserve, a 140-apartment eco-resort to provide leisure and cultural services for staff and their families.[96]

As part of its social remit, the reserve and resort also support people who might once have lived and worked on the ranches. The community at São Pedro de Joselândia, which is just outside its fence, gets some assistance. The resort also runs a butterfly house that is stocked with 3,000 butterflies raised from eggs by poor local women. The resort, which hosted us during our visit, introduced us to the women. One, Maria Laurinda de Carvalho, showed off her butterfly shed outside her home in the backstreets of Poconé. It was full of pots containing butterfly eggs, larvae and caterpillars at various stages of development. In the yard were plants that she grew to feed them. 'We take better care of the butterflies than of our husbands,' she joked.

It seemed a curious fate for former pantaneiros – banned from fishing in their rivers and corralled into towns such as Poconé, where they spend their days tending butterflies rather than cattle. But, Maria said, 'It gives me a reason to live. I was unemployed when I heard about the project.' She earns $250 a month and hopes to use it to train as a nurse.

While indigenous communities try to claw back their former territories, and conservation grabs more, much the largest category of land use in the Pantanal remains cattle ranching. More than 90 per cent of the Pantanal is within such estates, which extend over large areas and contain cattle at necessarily very low densities. Some ecologists see the ranches as potentially a more viable future for Pantanal ecosystems than conventional protected areas. The Macaulay Institute for land use research in Aberdeen, Scotland, for instance, has called ranching in the Pantanal 'one of the few examples of sustainable development introduced by Europeans into a tropical environment'.[97]

Down a wooded track 40 kilometres south of Poconé, we visited Luiz Vicente da Silva. He owns a ranch covering 70 square kilometres, which is small by comparison with many of his neighbours. He told us he believes cattle could be good for the Pantanal. Their grazing suppresses fire, he said. By comparison, protected areas that have excluded cattle quickly turned into tinderboxes. The SESC reserve, he noted, suffers around 5,000 fires a year, and employs twenty-five fire-fighters through the dry season to extinguish them. 'How sustainable is that?' he asked as we toured his fire-free spread.

Fire is a contentious issue in the Pantanal. It is natural here. Along with the floods, it drives the ecosystems, helping maintain a mosaic of grass and woodland in the drier areas. It promotes biodiversity. But you can have too much of a good thing. Without cattle or other grazing animals, the grasses in protected areas soon grow chest-high, creating the potential for major fires that 'destroy much more than the cattle do,' Vicente argued. 'I've seen it happen. In one fire near here, 300 giant anteaters died.'

What should be done? The practical solution to integrating these competing interests across the Pantanal may lie in a mix of indigenous territories, protected areas and grazing in ways that perpetuate the patchwork of ecosystems that make up the Pantanal.

If so, then Vicente is in the vanguard. While cattle are raised on part of his land, he has given the rest over to eco-tourists. His ranch is just off the Transpantaneira, a road into the Pantanal famous for having more than a hundred wooden bridges over wetland creeks.[98] Visitors can stay at his twenty-six-room lodge, surrounded by woodlands and swamps, and go looking for wildlife. 'I want to build a tower so visitors

A speeding tourist boat out at dawn in the southern Pantanal, looking for wildlife to view – or fish to catch.

can see the swamp better,' he said. On a track behind the lodge, we saw the footprint of a tapir in the dust and a caiman skull near a stream. One tree right by the lodge was a haven for macaws. 'We have three nesting rheas and several anacondas,' Vicente said, teasing Nunes about a visit years before, when she had waded chest-deep through his swamp with no concern for such giant reptiles.

Such creative compromises may have a bright future in the Pantanal. But any sustainable future for the wetland must above all address potential threats that lie outside its fluid borders. In particular, says Roque, the Pantanal is threatened by events in the savanna region immediately to the north, known as the cerrado. This plateau has become one of the hotspots of environmental destruction in Brazil. Its extensive grassland, which was once ranched much like the Pantanal, has been bought up by farmers to cultivate soy beans, corn, and cotton. For the decade to 2017, Brazilian Government data show that an estimated 140,000 square kilometres of native vegetation was cleared in the cerrado, twice the loss in the Amazon in the same period.[99]

This change matters to the Pantanal because the farms block animal species that once migrated between the two regions, and because most of the rivers that keep the Pantanal wet rise in the cerrado. Those rivers now bring agricultural chemicals into the Pantanal, and they are increasingly being blocked by dams. There are forty-five hydro-dams already, with more than a hundred planned. Most are little more than weirs. But not all. The 72-metre-high Manso Hydroelectric Dam has a reservoir covering more than 400 square kilometres. It holds back the wet-season flood on the River Cuiabá, the Pantanal's biggest river. Wetlands International has been working with local communities to identify ecologically important rivers that should be kept free from dams.

Meanwhile, the economic development of the cerrado is making the Pantanal much more accessible to Brazilians. New roads everywhere nibble at its edges, many of them advertised ominously as a 'gateway to the Pantanal'. The Transpantaneira stretches south from Poconé a third of the way across the wetland. For now, it is mainly used by tourists. But that would all change if the route were completed all the way to Corumbá.

The biggest threat of all, however, may be a direct assault on the hydrology of the Pantanal from the south. Most of the Pantanal's water ends up in the River Paraguay, which travels south into the River Paraná and enters the sea on the border between Uruguay and Argentina. It is one of the world's largest surviving free-flowing rivers.

But that would all change if the long-planned Hidrovia canal went ahead. The canal would turn the rivers Paraguay and Paraná and their marshy swamps into a shipping highway 50 metres wide and 4 metres deep. It would stretch for 3,400 kilometres all the way from the ocean to Cáceres on the north side of the Pantanal. The economic purpose of the canal is to turn the rivers into export routes for gypsum, iron ore, timber, and agricultural products from the heart of South America, including landlocked Paraguay and Bolivia.[100] But it would wreak hydrological and ecological havoc.

Hidrovia would open out the River Paraguay at two critical points in the Pantanal where the river narrows. Nicola, who campaigned against the project two decades ago, says the engineering involved would drain up to a third of the Pantanal.[101] The ecological impact would be huge, and the people downstream might not be too happy either. For the Pantanal currently functions like a large reservoir, storing water. Without it, floods downstream could be massive, she says.

In 2017, Wetlands International established a long-term programme to protect what it calls the Corredor Azul, or 'blue corridor', between the Pantanal and the ocean. This complex wetland corridor also includes the Iberá Marshes and the Paraná Estuary. Wetlands International, in collaboration with government agencies such as the National Wetlands Research Institute, wants to encourage a new outlook that would shun megaprojects such as Hidrovia in favour of more sustainable economic development.[102] In the Pantanal, this includes eco-tourism developments and combinations of farming and tourism, as well as assisting pantaneiros and indigenous communities such as the Kadiwéu to better access, manage, and benefit from their natural resources.

'One hundred and twenty-eight million people in South America depend on this system's health,' says Nicola. 'If rivers are dammed and tamed, benefits such as maintaining freshwater and food supplies for rural and urban areas will forever disappear.' That need not happen. But the task of charting an alternative path to development must begin in the Pantanal. o

RUPUNUNI, GUYANA
HOW TO MAP THE GLISTENING GRASSLANDS

NICHOLAS FREDERICKS IS A CHARISMATIC CHARACTER. A natural leader among the Wapichan, a tribe of some 9,000 people who occupy the grasslands of southern Guyana, a country on the northern shores of South America. He was brought up on his mother's cattle ranch, where he rode horses from a young age. Aged thirteen, he won the best-dressed cowboy prize at the annual rodeo in the region's only town, Lethem. He is now a father, and Chief of Shulinab, a community of some 500 people two hours' drive from Lethem, down a pot-holed and frequently flooded dirt road.[103] He is also in charge of documenting the Wapichan's claim to their forests and wetlands, which cover an area roughly the size of Wales.

The Wapichan territory lies deep inland, far from Guyana's populated coastal fringe. The expanse of bright green grass, occasional stunted trees and huge skies resembles East Africa. Marked on maps as Southern Rupununi, it borders Brazil to the west and south, and Surinam to the east. From May to August, it floods so extensively that early European explorers called it Lake Parime, and believed it might hide the fabled city of El Dorado.

The territory is a hydrological and ecological meeting place, where creeks that flow south into the Amazon mingle their waters with others that flow north to the Atlantic. There are jaguars, deer, ocelots, anacondas, caimans, tapirs, and armadillos.[104] The rich birdlife includes harpy eagles, the world's most powerful bird of prey, as well as pearl kites, savanna hawks, macaws, the near-endemic Finsch's euphonia, and the stand-out favourite, a small finch called the red siskin that was thought extinct until recently rediscovered by Wapichan rangers.

To an outsider, the lands of the Wapichan can appear empty of humanity. Unoccupied, even. Travelling between villages, we drove for hours without passing more than an occasional hunter or cattle herder. But the Wapichan use the abundant resources of their wet grassland with consummate care. They can spend weeks in the bush hunting for deer, bush hogs, agoutis, and armadillos with bows and arrows, or gathering medicines and traditional craft materials. But they only take what they need for their own families. Taking more, they say, would empty the larder for future generations. Their cattle range widely, but at densities too low to cause overgrazing. They farm in the forests, but their shifting cultivation leaves the land fallow for long periods before they return. The Wapichan leave some areas entirely alone because of their spiritual importance as burial grounds or former settlements, and sometimes as refuges for wildlife.

For much of the twentieth century, the Wapichan bled native trees for *balata*, a rubber-like latex that was one of the country's major exports until the 1970s. 'Men would spend months in the forest collecting balata,' said Tony James, an elder. 'Balata camps and trails could be found over almost the entire territory.' Traders built air strips in the bush to fly the latex out. But that has all ended. Only flying doctors use the air strips, and today the Wapichan probably engage in less outright commercial activity on their land than they have for a century. And that is proving a problem. The government in far-away Georgetown currently recognizes only around 15 per cent of the Wapichan's territory, and is handing out the rest to gold miners, loggers, and commercial ranchers.

During our visit, a Barbadian investor was reportedly creating a 12,000-hectare farm out of what he described as 'former wasteland', to grow rice, corn, and soy for export to Brazil. Another big agricultural company had its eyes on an area known locally as Machão-Pão, where the rare red siskins live. Illegal Brazilian gold miners regularly cross the border.[105] And the government had given a Chinese logging company access to almost 10,000 square kilometres of floodplain forest along the valleys of the Rewa and Essequibo rivers. Its workers were putting in a road to get the timber out. 'Our land is being taken from us often without us even knowing,' said Fredericks.

When the government asked them to justify their claim to wider areas of Southern Rupununi, the Wapichan took up the challenge. Fredericks organized volunteers equipped with GPS technology

The Wapichan people of the swamps of southern Guyana have combined traditional bushcraft with modern GPS technology to map their territories as part of a dispute over their rights to claim control of their traditional territories.

FOLLOWING PAGES:

Palm trees on the seasonally flooded savanna grasslands of southern Guyana – an area claimed by the Wapichan people.

on their phones to map and catalogue every feature, whether geographical, economic or cultural. The task took several years and involved most of the community. 'We assembled people in each village and found the elders and experts on the creeks, forests, and mountains. They were our guides as we walked, or took boats and bicycles and horses to survey the land,' said Angelbert Johnny, of Shawaraworo village, who helped organize the exercise. 'We went far from the villages, sometimes travelling for a month. It was hard. People got bitten by snakes. One guy had to be dragged out of the bush and given bush medicine. Another got lost for two days. But we went everywhere and mapped everything. Often we had to put our GPS equipment on poles and push them up through the forest canopy to get a signal.'

By the time the job was completed, they had logged 40,000 digital points. The maps were supplemented with detailed notes from elders

about the importance of every creek, homestead, and forest clearing. They had, Fredericks said, documented their land in much more detail, and with far greater accuracy, than the government ever had. But so far, their claim continues to be ignored by the government.

Fredericks longs for the day when the land politics is resolved and he can go back to his family's ranch. He admits ranching is no longer a thriving business in such a remote place. Getting the animals to market is too expensive. He sees nature tourism as an alternative. Currently, only a few hundred outsiders visit Southern Rupununi each year. Most are scientists, students, and eco-tourists, staying at the commercially owned Dadanawa Ranch. It once had more than 25,000 head of cattle, but is now devoted as much to conservation as to its cattle.

Wapichan villages would like a slice of that action. They have built a few small guest houses, in the hope of encouraging tourists to experience village life as well as wildlife. But, while the food is good, the accommodation rarely meets Western expectations. One guest house we stayed in required a 100-metre walk down a track in the darkness to find a latrine. There were rattlesnakes, we were told. Access to the villages remains poor, too. The road from Lethem is impassable during the wet season. We nearly got caught in the bush by a rising flood. But if tourists can get there, they will find a world that is a living refutation of the idea that wetlanders such as the Wapichan have to choose between modern and traditional ways. The next generation is wedded to both.

'Everything we want is here,' said 25-year-old Tessa Felix, who was both a bush tracker and her community's IT whizz. Back from a tour of villages across the Wapichan's flooded plains, she was trying to log into her email on a misty morning when the internet seemed slower than usual. 'My grandfather said that I was born from the earth and I believe that. Our land is like our mother. Now I want to fight for our land. My grandfather was a village chief. I want to continue his work, so we can govern the land which belongs to us.' She wants her traditional lands AND a better internet connection. ○

TONLE SAP, CAMBODIA
REVERSING RIVER IS THE BEATING HEART OF THE MEKONG

SOMETHING STIRRED BENEATH the boardwalk. There was a splash, and then another. Looking down, we had a surprise. Beneath our feet was a pen full of young crocodiles, feasting on a lunch of water snakes. We were in one of the sixty or so floating villages cruising the placid waters of what Cambodians call the Great Lake. These aquatic communities make their living from the lake and the flooded forests that surround it. This one had several hundred inhabitants, and probably rather more crocodiles. It was thriving and had an air of permanence, with vegetable gardens on pontoons, flowers at every door, and waterways lined with boat repair yards, filling stations, grocery stores, electricity generators, and even a karaoke bar. The farmed crocodiles sell for a thousand dollars apiece to Chinese traders. The people of the villages also fish, collect the eggs of the cormorants, pelicans, storks, and ibises, and net water snakes, frogs, reptiles, crabs, and molluscs.

The Great Lake owes its fecundity to the River Mekong, South-East Asia's greatest river, to which it is connected by a tributary of the main river known as the Tonle Sap. For much of the year the lake drains, through the Tonle Sap into the Mekong. But at the height of the annual monsoon flood, when the Mekong's flow is fifty times its dry-season level, the system goes into reverse. As the river's level rises by up to 9 metres, it backs up into the Tonle Sap. It is one of the wonders of the fluvial world: the river that goes into reverse. For around five months, from June to November, the Tonle Sap gushes upstream

LEFT: A baby is shaded in Kampong Phluk, a floating village on Cambodia's Great Lake. The lake fills during the annual monsoon as the engorged Tonle Sap river reverses its flow.

OPPOSITE: Fishers' houses on stilts in the flooded forests close to Angkor Wat, the great temple complex in the heart of Cambodia. The flooded forest nurtures millions of tonnes of fish. The flood is maintained by the wild waters of the River Mekong, and the fish migrate up and down the river each year. But dams threaten this natural cornucopia.

for around 120 kilometres into the Great Lake.[106] Above the town of Kampong Chhnang, the Great Lake spreads out into the surrounding forests, eventually covering some 16,000 square kilometres. It is, during this time, the largest freshwater lake in South-East Asia.

The reversing river carries with it fertile silt eroded from the Mekong's banks all the way from its sources on the Tibetan plateau. In the warm muddy lake waters, algae grow and feed trillions of fish fry, which grow into fat adults among the tree roots. More than 200 bird species feast here, including the endangered masked finfoot, and also long-tailed macaques, as well as the rare hairy-nosed otter. It is one of the most productive ecosystems on Earth.[107]

Humans have long exploited these wetland riches. Centuries ago, the Great Lake fed one of the world's largest pre-industrial urban societies, centred on the temple complex of Angkor Wat, which was built on its shores. At the height of its power in the twelfth and thirteenth centuries, Angkor ruled an area from the uplands of Laos to the Malay Peninsula. The temples, which are now major tourist attractions, were surrounded by a hinterland of roads, canals, and reservoirs stretching across the floodplain. There is little doubt how its people were fed: the temples' walls are covered in images of fishing.[108] Nature's bounty is undiminished today. The rights to fish much of the lake and flooded forest are auctioned for immense sums, with Vietnamese migrants being prominent buyers.

But just as the Mekong maintains the natural wealth of the Great

Lake, so the Great Lake sustains the life of the Mekong. As the Great Lake empties each November, fattened fish nurtured in the flooded forests head for the Mekong. Fishers down the Tonle Sap set up fences to catch them. It is a joyful time. Ever since Angkor's heyday, the people of the Tonle Sap have marked this annual bonanza with a water festival where the Tonle Sap enters the Mekong, in the modern Cambodian capital of Phnom Penh. We joined in, watching as hundreds of racing boats, decorated with images of water serpents, were paddled ferociously down the river to a finishing line at the Royal Palace, where a million or more Cambodians watched from the river banks.

The cornucopia extends far beyond the lake and the Tonle Sap. Many of the fish elude the fishers, enter the Mekong and migrate for hundreds of kilometres up and down the river. Two-thirds of the fish in the Mekong begin their life in the Tonle Sap – filling nets that feed tens of millions of people in Cambodia, Vietnam, Laos, and Thailand. More than 2 million tonnes are taken from the river each year. According to the UN Food and Agriculture Organization, river fisheries in Cambodia 'contribute more to the national food balance than any other inland fishery in the world'.[109] They provide three-quarters of Cambodians' protein intake, making some of the poorest people in Asia among the best fed.

Is all this harvesting too much? Some fear so. Certainly, the wild Siamese crocodiles are now largely extinct in the Great Lake, though the floating pens may house more than ever swam in the open water. Many fishers on the lake complain that their catches are not what they were. But fishery scientists caution against demonizing the fishers. They argue that the fishers are the true friends of the lake's wildlife. Their occupation of the flooded forests stops farmers from invading to make room for rice fields, which would destroy the entire system.[110]

But the big concern today is whether the great monsoon flood on the Mekong, one of the wonders of the fluvial world, will be allowed to continue. For, upstream on the river, dams are being built all the way from Cambodia, through Laos and Thailand, to China where it is called the Lancang. Around sixty dams are already in operation, mostly on tributaries in the lower basin.[108,111] The first on the main stem of the lower river, the Xayaburi Dam in Laos, was completed in 2019. More are planned. But the biggest worry lies in China, which is building a cascade of six huge dams on the main river. One, the Xiaowan Dam, is as high as the Eiffel Tower. Most of the dams generate hydroelectricity. To do that, they hold back the monsoon floodwaters and deliver downstream a steady flow to meet the needs of the region's electricity grids. That may be good for keeping the lights on from Shanghai to Phnom Penh, but it reduces the monsoon flood pulse that is the key to the Mekong's stupendously productive ecosystem.

China said in 2018 that the operation of these dams had reduced river flow out of China by 30 per cent in the flood season, and increased it by 70 per cent in the dry season. The effects on water flow further downstream are smaller because, for now at least, tributaries joining the main river help maintain the monsoon flood. The Tonle Sap continues to reverse. But the Chinese dams have a much bigger effect on silt flowing downstream. Until the dams' construction, around half the silt in the Mekong arose in the Chinese stretch. But

PREVIOUS PAGES:
Fishing rights in the Great Lake at Angkor Wat are highly prized. The catches are greatest when the fish migrate back to the River Mekong at the end of the monsoon season.

now most of that silt is caught behind the dams. Just upstream of the junction with the Tonle Sap, suspended sediment in the river has fallen by more than 40 per cent.[112] The Great Lake is being starved of silt, and there are grave fears that this will undermine the fertility of the lake.

China has claimed that the reduced flood pulse is beneficial for its downstream neighbours because it smooths out river flow, reducing the risk of floods and droughts. But in Cambodia, fishers see things differently. For them, the monsoon floods are good. They bring the water and silt that washes up the Tonle Sap into the Great Lake, and create the lake's biological bonanza. Suppressing the flood pulse could one day end the Tonle Sap's miraculous reversal, and the lake's cornucopia would be over. Even if the pulse persisted, the collapse of silt supplies will almost certainly crash the productivity of the Great Lake, and with it the fisheries of the wider Mekong.[113,114]

There is no inevitability about this, however. After years of refusing to discuss the operation of its new dams, China has entered into talks with its downstream neighbours about the consequences for the Mekong. Chinese leaders now stress the need for an 'eco-civilization' at home. To ensure the same for its neighbours, the maintenance of the monsoon flood, silt supplies, and the annual reversal of the Tonle Sap should be top of the agenda.

THE MEKONG SEASONAL FLOOD PULSE is a marvel because it has survived. Dozens of other great river systems were once as productive as the Mekong is today. But dams have drastically diminished their wetlands and crippled their freshwater ecosystems. Around the world, there are almost 60,000 large dams. More than 20,000 of them are in China; the US, India, Japan, Brazil, South Korea, Canada, South Africa, Spain, and Albania all have more than a thousand each.[115] The dams mostly either directly abstract water for irrigation and human settlements, or store it for discharge downstream through turbines to generate electricity. Thus, they either empty rivers or suppress the seasonal floods that sustain floodplain wetlands downstream. According to one detailed study, led by Gunther Grill of McGill University, Montreal, roughly half the world's total river flow is 'moderately to severely impacted' by dams, a figure that he says could rise to more than 90 per cent if all planned dams are completed.[116]

Luckily, knowledge about the downsides of dams is growing. Especially in North America and Western Europe, this has resulted in a drastic decline in dam building. Nowadays, some countries are taking down more dams than they are raising. But the construction companies have moved to the developing world, where the same problems are being repeated. The arid regions of the Sahel and East Africa in particular are undergoing an orgy of dam building, even though many of the projects directly threaten the functioning of vital wetlands.

As we saw earlier, Ethiopia's Gibe III Dam on the River Omo was, in 2018, emptying Lake Turkana in Kenya. In the same year, the Kenyan Government announced the go-ahead for the High Grand Falls Dam on the country's longest river, the River Tana, just upstream of its extensive and highly productive delta; and Tanzania let contracts to build a dam that would stem silt input to the Rufiji Delta. Meanwhile, in West Africa, Guinea said it would go ahead with the long-planned Fomi Dam on the River Niger, directly threatening the Inner Niger Delta, an ecological jewel on the edge of the Sahara Desert.[117]

Dams have their defenders. Governments often say their countries urgently need them to provide water for cities and irrigation, and to harness hydropower. As Ethiopian Prime Minister Meles Zenawi put it in support of his Gibe III Dam in 2011, 'We cannot afford not to have Gibe III.'[118] ... We want our people to have a modern life.'[119] The dam's promoters often add that hydroelectricity is a low-carbon source of energy.

Such arguments have some validity. But the downstream economic harm caused by disrupted river flows, lost silt and damaged wetlands is rarely assessed with the alacrity of the upstream benefits. And even the latter benefits are often hyped by their promoters. Oxford geographer Atif Ansar reported in 2014 that as many as half of the large dams built in the past century have had a negative economic return.[120] The dams studied came in on average at twice the prospective budget. There had been no improvement with time, and the bigger the dam, the worse the outcome.

Another review in 2017 found that large dams that promise to improve water supplies usually create more water scarcity than plenty. Ted Veldkamp of the Free University, Amsterdam found that 23 per cent of the world's population has been left with less water as a result of dam construction, compared to only 20 per cent who have gained.[121] Winners were mostly upstream, and were outnumbered by downstream losers. Certainly, on the Mekong there can be little doubt that however many upstream winners there are from the planned dams, those who depend on the abundant free resources of the Mekong's downstream floodplain are set to be losers. ○

LAKE CHAD, WEST AFRICA
DAMS, DYKES AND AN INTERNATIONAL REFUGEE CRISIS

A LMOST THIRTY YEARS AGO, in the desert margins of northern Nigeria, a hundred or more youths, wearing bright pink shorts and brandishing fishing nets, stood poised beside a river. At a signal, they all plunged into the water. A mass of flailing nets and bodies fought for fish trapped as the river emptied during the long dry season. It was the start of the Gorgoram Fishing Festival, a celebration of nature's bounty held every February in the Hadejia-Nguru Wetland, a large green oasis on the edge of the Sahara Desert.

Half an hour after taking to the water, the glistening young men were carrying their catches up the banks to a huge pair of scales erected in front of a grandstand. Three local emirs and a national presidential candidate watched as they weighed the whiskered catfish, still twisting in the midday sun. There were prizes for the best hauls – a bicycle for one village team, a mechanical sewing machine for another. But it was

clear that the festival, the high spot of the year on one of Nigeria's largest wetlands, was a shadow of its former self. The scales could accommodate individual fish weighing up to 100 kilograms. But the largest fish caught that day managed just 4 kilograms. Something had gone wrong on the wetland.

The prizes were duly awarded, but the youths looked disappointed and resigned. The girls watching from the grandstand looked cheated. There were no heroes that day. The crowd started to drift away. The river and its fish were forgotten. But the ecological crisis on the wetland was all around. As the convoy of departing dignitaries bounced down the long dirt track that wound past one of the emirs' palaces to the nearest tarred road, clouds of dust rose on the desert wind. The land was littered with fallen trees, victims of a sinking water table.

The scene at the Gorgoram Fishing Festival was played out in early 1992. Back then, there was still some life in the wetland. Herds of cattle waded through its waters to graze on lush grasses. Birds flew in from Europe to winter on the diminished lakes. Farmers planted rice and millet as the waters receded from seasonally flooded soils. But in the hills behind, engineers had erected dams that captured 80 per cent of the flow of the two rivers that had once nurtured a wetland stretching for more than 100 kilometres that was the driving force behind almost every aspect of the local economy. The wetland was contracting year on year. Its demise was only a matter of time.

One of us (FP) wrote at the time, 'It is not, for the moment, a disaster that makes headlines. There are no starving children, no airlifts of grain, no refugee camps. But I may have been looking at the start of a silent, stealthy, ecological apocalypse that a decade from now might grab the world's attention. By then it will be too late to help.'[122] And so, sadly, it has proved.

The fishing festival struggled on for a few more years. The last was held in 1998.[123] There simply wasn't enough water to sustain the event. Around four-fifths of the wetland has now dried up. Camels walk where fishermen once threw their nets. And without water percolating underground from the wetland, the water table has fallen by as much as 25 metres, killing many more trees and drying up wells in dozens of villages. There are annual battles over access to water, between the permanent residents and the nomadic cattle-rearing Fulani, who depend on the wetland to graze their animals during the dry season.

The dams were built to provide water for Kano, the biggest city in northern Nigeria, and to irrigate commercial crops. Politicians promised that they would bring wealth to a parched region. It is true that a few well-connected people with fields irrigated by the dams have since prospered, growing commercial crops such as wheat sold for making bread in Lagos or Abuja. But the thousands of herders, fishers, and farmers who used the wetland have lost out. An economic balance sheet drawn up by Edward Barbier of Colorado State University presents estimates suggesting that the overall economic losses may have been six times greater than the gains.[124]

The consequences of this disaster have reverberated far beyond the Hadejia-Nguru Wetland. Many of the people who lost their

An army officer from Chad patrols what remains of Lake Chad. Thanks largely to dams on the rivers that fill the lake, it is usually only around a tenth of its former size and, as shown here, is now largely covered in vegetation.

RIVER FLOODPLAINS

livelihoods there will have joined the Islamic terrorist group Boko Haram that grew strong in this part of north-east Nigeria during the 2010s. Perhaps the children of those glistening youths were among them. The area has become a haven for suicide bombers, kidnappers of schoolgirls, and murderers. Many people have given up and left, paying people-smugglers to take them to Europe, part of an influx that has triggered a crisis of governments in the European Union.[125,126] Too late, the world's attention has been grabbed.

THE COLLAPSE OF THE HADEJIA-NGURU WETLAND is the best-studied part of a wider hydrological crisis. Until half a century ago, the rivers that fed the wetland carried on east. They helped fill Lake Chad, then Africa's fourth largest lake, straddling the borders between four countries: Nigeria, Niger, Chad, and Cameroon. But since then, starved of water from upstream, the lake has lost more than 90 per cent of its surface area. Only Chad and Cameroon still have lake shorelines.[127] Initially, drought in the Sahel helped drive the lake's decline. But since 2002, rainfall has improved markedly, while Lake Chad has failed to recover because new dams on the rivers flowing into the lake now divert water away from the lake.[128] As the Hadejia-Nguru Wetland has dried, water inputs from Nigeria to Lake Chad have almost entirely dried up. In Cameroon, the other main source of water for Lake Chad, the River Logone, has suffered similarly.

Since 1979, the Maga Dam has blocked the River Logone. Before that, the river's wet-season flow equalled that of the Rhine in Germany, and 6,000 square kilometres of floodplain pastures benefited. But the 30-kilometre earth embankment now diverts two-thirds of the river's waters to rice farms, and most of the floodplain is dry.[129] The UN Environment Programme estimates that 20,000 cattle lost their pastures because of the dam; fish catches fell by 90 per cent; and village harvests of sorghum, millet, and rice declined by 75 per cent. Wells and water holes emptied. Up to 40 per cent of the human population left. Wildlife joined them. As vegetation died, kob antelopes and other animals fled the Waza National Park, which occupies part of the floodplain.[130]

As on the Hadejia-Nguru Wetland, the losses brought few compensating economic benefits.[129] Most years, little rice grew on the new irrigated rice fields. Only around a tenth of the water collected by the dam was productively used. Overall, the dam and its works 'diminished rather than improved the living standards and economy of the region,' concluded Paul Loth of Leiden University.[131,132] 'A once

What remains of Lake Chad: a composite satellite image taken in 2019.

fertile floodplain ... turned into a dust bowl.' Loth proposed bringing back the flood. For a while the Cameroon Government listened. It allowed a trial reflooding in the 1990s.[130] During the trial, cattle numbers rose by 260 per cent, without any sign of damage to the grasslands.[132] Fish returned. Wells revived. Loth published a book called *The Return of the Water*.[132] It was a good news story about how to revive a wetland. But it proved to be a flash in the pan. After a couple of years, the government ended the trial. It has never been revived. But it still could be.

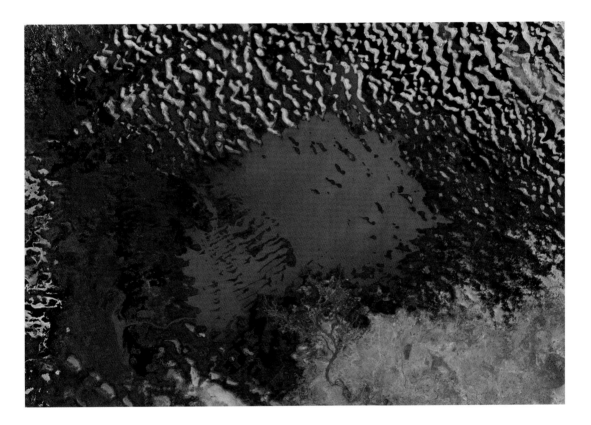

The loss of the water flowing through wetlands such as Hadejia-Nguru and the Logone floodplain has all but killed Lake Chad, though it would quickly recover if the waters returned. Most years, it is less than 10 per cent of its former size, and much of what remains is covered in weed. Social breakdown and conflict have followed in its wake. Old animosities between herders and farmers over access to vital lands and water have flared up. In 2016, 2,500 people were killed and 62,000 people displaced by such conflicts in northern Nigeria.[133] Hundreds more died in early 2018. In a climate of escalating violence, more than 2.6 million people left the Lake Chad region between mid-

2013 and mid-2016, according to the UN's International Organization for Migration.¹³⁴

The links in the causal chain from water management to wetland health to social breakdown and international migration are complex. Wetland loss is certainly not the only reason for the breakdown of law and order and the human exodus from the Sahel. And migration is a long-standing coping strategy in a region of extreme climate variability. But the parlous state of the wetlands is changing the region. In the past, wetlands were refuges during drought or conflict. They were safe and fertile places to hide. But today, with their waters diminished, they have become places of shortages, hotbeds of conflict, and sources of outward migration. Moreover, migrations that were once temporary and local are becoming permanent and intercontinental.

Slowly, the agencies that helped cause the hydrological disaster are recognizing their folly. In 2015, Mana Boukary of the Lake Chad Basin Commission, an intergovernmental body that has presided silently over the lake's decline, said, 'Youths in the Lake Chad Basin are joining Boko Haram because of lack of jobs and difficult economic conditions resulting from the drying up of the lake.'¹³⁵ The UN humanitarian coordinator Toby Lanzer told a summit between the European Union and African leaders that the dislocation was also fuelling migration. 'Asylum seeking, the refugee crisis, the environmental crisis, the instability that extremists sow – all of those issues converge in the Lake Chad Basin.'¹³⁶ Even an audit of the lake basin by the Nigerian Government, the sponsor of some of the biggest dams depleting the lake, conceded that 'uncoordinated upstream water impounding and withdrawal' were among factors that had 'created high competition for scarce water, resulting into [sic] conflicts and forced migration'.¹²⁴

But this new-found realism has not extended to finding realistic solutions. The Nigerian Government has instead promoted a fanciful idea to refill Lake Chad with water brought by digging a canal all the way from the River Ubangi in the Congo Basin, some 2,400 kilometres to the south. In 2017, the cost of the scheme was put at an eye-watering $50 billion.¹³⁷,¹³⁸ It has no realistic chance of being built. Even if it were, it would cause huge ecological disruption of its own. Such distractions merely prevent sensible debates about how to use existing water better, and how to restore the natural bounty once provided by wetlands on the edge of the Sahara. There still seems little appetite for that.

Fish are scarce in Lake Chad today. The once-mighty inland sea is now only a tenth of its former size. Dams built to irrigate crops dry up many of the rivers that once maintained its waters. Ecological decline has led to social breakdown, terrorism, and outward migration, including to Europe.

THIS CHAIN OF ECOLOGICAL DECLINE and social implosion in the Sahel extends beyond the Lake Chad Basin. Farmers, herders, and fishers along the valley of the River Senegal, which forms the border between Senegal and Mauritania in West Africa, told us of their struggles to make ends meet since the completion in 1988 of the Manantali Dam upstream in Mali. Before its flow was blocked, the river regularly watered an extensive floodplain. Millions of people relied on its waters to provide fish, to wet soils for crops, and to nourish grass for pastoralists. Now, the dam holds back the seasonal flood. It generates hydroelectricity for cities and delivers irrigation water for commercial agriculturalists growing cash crops such as sugar, cotton, and rice. But its management has reduced the river's fisheries by about 90 per cent, and reduced the area of the floodplain benefiting from its annual flood by up to 2,500 square kilometres.[139]

Seydou Ibrahima Ly, a teacher in the bankside village of Donaye Taredji in Senegal's Podor district, told us during our tour of the area in 2016 that, when he was young, 'the river had a flood that watered wetlands where fish grew.' But 'now there is no flood because of the dam. … Compared to the past, there aren't many fish. Our grandparents did a lot of fishing, but we don't.' People have been fleeing the desiccated floodplain, he said. More than 100 people had left his small village alone. 'In some villages, they are almost all gone.'[140]

Many head for Europe. They know well enough that the boats across the Mediterranean are dangerous, said Oumar Ciré Ly, Deputy Chief of neighbouring Donaye village, which has also seen an exodus of its young people. 'But they have a determination to find a better life.' Some succeed. Oumar's brother is now a lecturer at the University of Le Havre in France. 'He still sends back money to the village.' Families left behind on the dried-up floodplain often survive on such remittances, he said.

The Senegal River Basin Development Organization – the intergovernmental agency that built the Manantali Dam – conceded as long ago as 2007 that eliminating the river's annual flood 'has made flood-recession crops and fishing on the floodplain more precarious, which makes the rural production systems of the middle valley less diversified, and therefore more vulnerable'.[141] But it seems not to care. Amadou Lamine Ndiaye, its Director of Environment and Sustainable Development, told us that his agency regarded wetlands such as river floodplains primarily as a resource for wildlife that brought in revenue from tourists, rather than as a lifeline for rural communities. So, wetlands where foreign visitors come to admire the birdlife should be protected, but the rest did not merit the same attention. That attitude, surely, must change. o

INNER NIGER DELTA, MALI
DESERT JEWEL ON THE BRINK

MANY RIVERS FORM A DELTA as they enter the ocean. But some also form deltas when they hit flat plains far inland. The Niger, West Africa's largest river, spreads its waters across desert near Timbuktu in Mali, an ancient centre of trade and learning that it once sustained. Today, around two million herders, farmers, and fishers from different ethnic groups live in the Inner Niger Delta, among its complex network of meandering channels, swamps and lakes, which covers an area the size of Belgium.[142]

The delta changes through the year. During the wet season, it is inundated and the only means of travel is by boat. It is full of fish that attract water birds all the way from Europe. Kingfishers, marsh harriers, purple herons, and cormorants abound. Fishers from the Bozo ethnic group punt, sail, and row their pirogues through the delta, setting up camps far from home as they follow their quarry. Once the waters recede, Bambara farmers, descendants of the founders of the Malian Empire at Timbuktu, come to the fore, planting rice, millet, and other crops in the wet mud. Then Fulani herders arrive from as far away as Mauritania and Burkina Faso to graze their cattle on the rich hippo grasses (also known as *bourgou*) that are exposed by the retreating water. After harvest time, as the surrounding landscape turns to dry pasture that is navigable only by 4x4 vehicles, the animals nibble at the field stubble.

The three groups have long-established rules for sharing the

wetland's resources. There are tensions, of course. Fishers like to keep cattle out of the *bourgou*, because they know they get bigger fish if they do. But they also know that as the waters retreat, cattle trample broken grass stems into the mud, which helps propagate the bourgou and stimulate fish breeding for next year. Timing is everything. 'We tell the herders they can only go [into the bourgou] when the big fish have been caught,' said one Bozo fisher. Relations are amicable, he insisted. 'Bozo and Fulani have lived together here for centuries.'

Mali is one of Africa's poorest countries, but the delta provides 80 per cent of its fish. It has some of the highest livestock densities in

Africa, nourishing two million cattle – a third of Mali's stock – as well as three million sheep and goats.[143] It delivers 8 per cent of GDP and directly feeds 14 per cent of the country's population.[144] In fact, there is spare food for export. We watched trucks leaving Mopti, a market town on the southern shore of the delta, to deliver delta fish, packed in crates with ice, across West Africa. The journey to Nigeria can take a week, one driver said.

On our journeys across the delta, we found that the main preoccupation of most villages was managing the wetland to make best use of its water. With more people and less water coming down the river into the delta, new ideas are vital. Several communities we visited had taken to channelling the silty delta water into fish ponds, onto grazing pastures, or down furrows to irrigate crops through the dry season. We visited Noga, near the banks of one of the main river channels. The channel used to overflow its banks and water the villagers' fields in the wet season. But now it rarely reaches the top of the banks. So, they have dug a hole in the bank and carefully direct the precious water through stands of mimosa bushes and down a cascade of five newly dug fish ponds, the biggest covering 20 hectares. It was the dry season when we visited, but village secretary Baba Troulet said, 'The ponds should flood for five months. They will be full of fish, as well as birds like egrets and herons, and grasses that the fish will eat. Then, the water will overflow into the bush so we will have more pasture.'

OPPOSITE: Cattle graze the *bourgou* grasses of Mali's Inner Niger Delta – a superbly productive wetland on the edge of the Sahara desert – but threatened by dams upstream.

RIGHT: Women in the Inner Niger Delta selecting fish from the day's catch. But catches have reduced because the wetland is being diminished by upstream water abstraction.

FOLLOWING PAGES:
At the height of the wet season, most of the Inner Niger Delta is flooded and travel is by boat. But in the dry season the water gives way to desert. You need a 4x4 vehicle here.

RIVER FLOODPLAINS

Nearby, Simina village has dug a dyke more than a kilometre long to divert water to a fish pond that doubles as a reservoir for cattle and farming. 'The dyke is our future,' said Yousseff Traoré, the village secretary. 'Without it, future generations will move away and the village will die.'

Fishers are innovating, too. We spoke to some in Akka, a village on the northern shore of Lake Débo, the largest lake in the delta. They have begun planting bourgou in the lake's shallows to encourage more fish. The 30-hectare stand is so successful that the shoals of fish attract thousands of cormorants and pelicans. But the villagers don't worry about sharing the fish with the birds. The birds' droppings fertilize the water and mean there will soon be more bourgou, they said.

Meanwhile, women have banded together to irrigate vegetable gardens and plant fruit trees. This is almost revolutionary, one group said. Traditionally in Mali, women do not own farmland. But the men recognize the extended kitchen gardens as theirs. In Kakagna, two groups of women each plant onions, lettuce, tomatoes, okra, and cabbages that feed their families and earn income at local markets. 'The men are happy too, because women now contribute two-thirds of the family income,' said Diko Bilakoro, the President of one of the groups. Though in Noga the men complained that the vegetables had been planted on their soccer pitch.

Many of these small projects have been supported by Wetlands International with technical expertise and micro-credit.[145] By and large, women's groups run the cash side of things in the villages, often topping up the initial credit with small subscriptions of their own. We saw this in action, with women bringing the cash to village meetings padlocked in metal boxes. At Gouraou Bozo, we asked the women why men were excluded. It was women's business, they said. 'Anyway, if the men were involved, they would use the money to marry another lady.'[1]

YOU WOULD THINK THE GOVERNMENT of Mali would be keen to maintain this highly sustainable way of living on the edge of the Sahara, and to encourage the wetlanders' innovation. Yet the delta often seems left out of government economic plans devised in the capital, Bamako, 500 kilometres to the south. We found few efforts under way to help the wetlanders use their water better. Instead, the very future of the delta is threatened by schemes dreamed up in Bamako to capture its water for economic purposes deemed to have a higher priority.

In recent years, the Malian Government has dammed and diverted water from the River Niger just upstream of the delta, to

As the waters recede in the Inner Niger Delta, farmers plant crops in the mud. These women are planting vegetables such as okra to improve the diets of their families. Wetlands International has helped them channel water to establish their vegetable gardens.

irrigate desert fields of thirsty crops such as rice and cotton, often established by foreign investors. These diversions have cut the area of delta flooded by up to 7 per cent, causing declines in wetland trees, fisheries, and grazing grasses, says Dutch hydrologist Leo Zwarts.[146] The northern part of the delta around Lake Faquibine, which was once central to the wealth of Timbuktu, is now usually dry.

The water shortages have exacerbated tensions between the communities on the delta, with some among the Fulani, in particular, radicalized by jihadist groups from outside, since the civil war in 2012, under the banner of the Macina Liberation Front, named after an empire that ruled here two centuries ago.[147]

Faced with water shortages and rising community conflict, some people have left the delta. Some have been among the Malians regularly reported to be filling migrant boats heading from Libya to Italy.[148] This trickle could become a flood if Mali's upstream neighbour, Guinea, completes a giant new hydroelectric dam in the River Niger's headwaters.[149] In 2017, Guinea announced an agreement with Chinese construction companies to build the Fomi Dam.[148] To keep its turbines humming year-round, the dam is likely to replace the annual flood downstream towards the wetland with a more regular flow. That regular flow will also allow the Malian Government to tap much more of the river's water for its desert farms, fulfilling a long-standing plan to triple the area of land under irrigation along the river.

If all that happens, the amount of water reaching the wetland will decline by about a quarter; and it will arrive in a regular flow rather than a flood pulse. Zwarts estimates that the altered hydrology could reduce fish catches in the delta by 30 per cent, and diminish its wet pastures by almost as much.[150] In effect, he says, drought conditions such as those last experienced in 1984, when three-quarters of the wetland dried out and people fled en masse, would become the new hydrological normal. One of the world's greatest and most productive wetlands will face disaster.

But there is nothing inevitable about this. Even if the dam is built, it could be operated to create an artificial flood downstream. And the Malian Government could ensure that water reaches the delta by scaling back its plans to expand irrigated agriculture in the desert. Studies carried out by Wetlands International and its partners have established what river flows are necessary to sustain and restore the delta and its biodiversity, while optimizing productivity, as part of an overall development plan for the delta and the Upper Niger River Basin. European governments who are intent on limiting migration from the region across the Mediterranean might wish to encourage such an approach, while providing assistance for communities in the delta to make the most of their wetland ecosystem. ○

SUDD SWAMP AND MESOPOTAMIAN MARSHES
REFUGES IN TIMES OF CONFLICT

OUTSIDERS HAVE LONG FEARED and loathed the Sudd. The British explorer Sir Samuel White Baker called it 'this horrible region of everlasting swamp'.[151] Africa's second biggest wetland was a barrier to British schemes to make the White Nile navigable, and to bring the river under imperial control all the way from Egypt to Lake Victoria. Also, it was a waste of water. They estimated that half the water that meandered slowly through its myriad channels, around 5 cubic kilometres a year, was lost to evaporation in the fierce desert sun. In 1904, they drew up a plan to bypass the Sudd with a canal through the Jonglei Desert to the east. That way, ships could push upstream unhindered, and the Nile's waters would avoid evaporation and quickly flow downstream to irrigate farms in Sudan and Egypt. Deprived of water, the Sudd would wither away.

Well, it hasn't happened. The European imperial era ended with the Sudd still swelling to more than 40,000 square kilometres in the wet season.[152] Its fish stocks are largely undiminished, despite as many as a million people spending at least part of the year in and around its waters. And more than a million antelopes – mostly horned tiangs and white-eared kobs – still migrate between the Sudd and the flooded plains of Gambella in Ethiopia, in a migration second only to the wildebeests of the Serengeti.

It was a close-run thing, however. In 1978, after decades of delay, the Sudanese authorities began to cut the Jonglei Canal. They brought to the edge of the Sudd a giant digging machine, five storeys high, weighing 2,300 tonnes, and known as the bucket-wheel. There was opposition to the canal, especially from the local Shilluk fishers and Dinka herdsmen, who brought half a million cattle to graze in the wetland in the dry season. Even if the canal did not entirely dry up the Sudd, its 50-metre-wide channel would prevent their animals from reaching their pastures. But in the Sudanese capital of Khartoum, Vice-President Abel Alier declared, 'If we have to drive our people to paradise with sticks, we will do so, for their own good and the good of those who come after us.'[153]

That was the plan, but the Dinka have so far had the last word. They got up a rebellion. Rebels with bases in the Sudd killed a pilot servicing the project, and kidnapped several of the engineers. The rest fled and the giant bucket-wheel was abandoned, its job only half done. It still remains where they left it, slowly decaying in the desert. The rebels continued a wider fight for the independence of southern Sudan. In 2011, after three decades of fighting, they declared an independent state of South Sudan. Khartoum rule was finally banished from the Sudd.

But the politics has moved on. There are new rebels in the Sudd. South Sudan has been engulfed by civil war and in mid-2018, thousands took refuge in the swamps after the government set fire to their huts and shelled their villages. At the peak of fighting, the UN estimated that as many as 100,000 people were living there.[154] In Juba, South Sudan's capital, some now say that resurrecting the bucket-wheel will raise money for a beleaguered government by selling Egypt the water it desires, while also flushing out the rebels.

Maybe. But for now, the Sudd has had the last word. Its wildlife has survived the decades of conflict. Arguably the fighting provided protection. Would-be bush meat hunters and ivory traders have been for once outgunned by those hunting down humans. The fish eagle,

The Sudd wetland is a rich resource for wildlife as well as humans. Here, elephants migrate through the Shambe National Park.

FOLLOWING PAGES:

The Sudd, on the White Nile in South Sudan, is one of Africa's largest wetlands. These villages, at the southern end of the swamp near the town of Bor, have become refuges during decades of conflict.

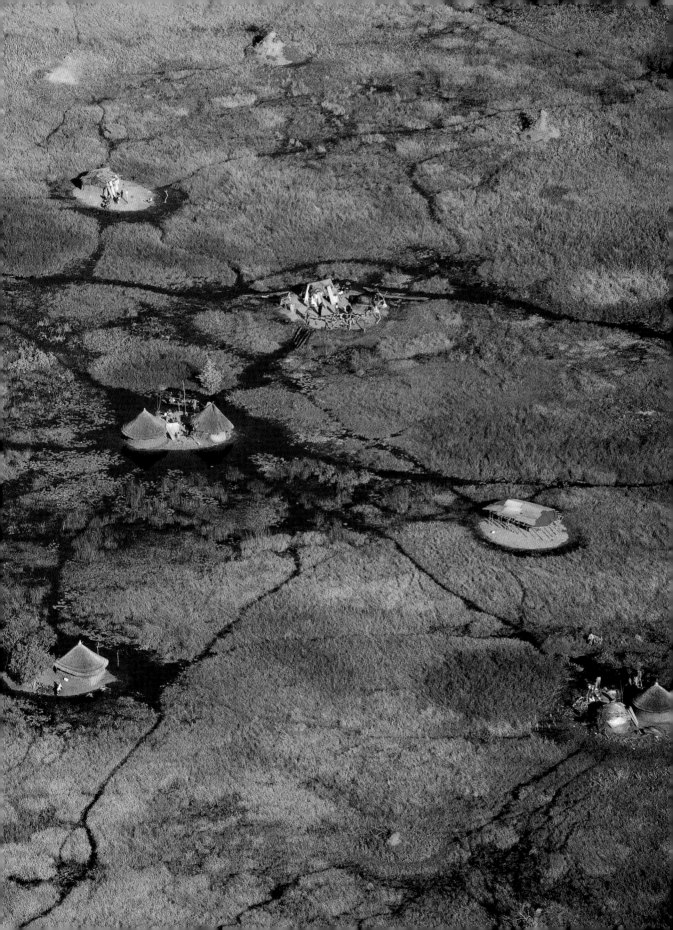

South Sudan's national bird, still hunts there. The great wetland refuge in the heart of Africa remains just that.

UNFENCED WETLANDS ARE OFTEN legally unowned and open to all. To some that makes them lawless badlands. But, as commonly owned natural resources, they are also vital in hard times, whether during droughts or wars and social unrest. Often remote, they also lend themselves to people with a self-sufficient lifestyle and culture, and to those marginalized in society. To outsiders, 'taming' such inhabitants

can seem of a piece with 'taming' their waters. But trying to drain the swamps can exacerbate rather than suppress conflicts, and create new forms of social discord. It has happened in the Sudd. And so, too, in the marshes of southern Iraq.

The Mesopotamian Marshes have for thousands of years been the largest wetland in the Middle East. Their huge expanses of reed beds have for 6,000 years been the home of the Ma'dan people, often called the Marsh Arabs. The people and the wetland have a symbiotic relationship. The marshes provide almost everything the Ma'dan want. Ecologists say that it is only their annual burning of the reed beds, and the constant grazing of their water buffalo, that have kept the marshes from being choked with the vegetation, filling with mud and drying up.[155]

The marshes are sometimes regarded as the birthplace of the story of the Garden of Eden. But they offer a modern fable too – of conflict, loss and possible redemption.

As late as the 1980s, the marshes extended for up to 20,000 square kilometres across the floodplains of the Tigris and Euphrates rivers, south of Baghdad. The Ma'dan way of life persisted, even though by now the marshes supplied two-thirds of Iraq's fish. But all that was shattered in the 1990s by Iraq's then President Saddam Hussein. In the aftermath of the First Gulf War, when the Ma'dan joined Iran-backed opposition to his rule, Saddam decided to besiege them by blocking the two rivers. He followed an engineering blueprint developed by the British in the 1950s to capture the marsh's waters and develop irrigated agriculture on the drained land.[156,157] But he never built the irrigation works, because his prime aim was to starve the Marsh Arabs out of their homeland.

His brutal act of ecological warfare succeeded. Two of the three main marsh areas – the Central and Hammar Marshes – almost entirely dried up. 'One of the world's largest and most significant wetland ecosystems has completely collapsed,' the UN Environment Programme reported in 2001.[158] It compared the catastrophe to the desiccation of the Aral Sea and the deforestation of the Amazon. The Ma'dan went into exile in Iran. But obliterating the marshes had an unintended consequence for Saddam. A decade later, when American tanks invaded Iraq from the south during the Second Gulf War, they rolled in over former marshes that a decade earlier would have consumed them.

With Saddam overthrown, the Ma'dan returned to their former homeland. They broke his dykes and reflooded the marshes. A new government in Baghdad helped them by constructing banks and canals to keep parts of the marshes wet all year round. By 2013, three-quarters of the marshes were again flooded. Conservationists reported that all 178 bird species previously recorded there had returned, and the government formally declared the marshes a national park. It was a remarkable story of ecological redemption. Testimony to the ability of a wetland to recolonize once its water returns; reinforced when UNESCO declared the marshes a World Heritage Site for both its cultural and natural heritage.[159] Wetlands International is working with the Iraqi Government on a long-term management plan to protect nature and local livelihoods.

The Ma'dan people, or Marsh Arabs, lived traditional lives in southern Iraq's Mesopotamian Marshes – until forced out by Saddam Hussein, who diverted the rivers that watered the marshes. After Saddam's overthrow they returned, but their old lifestyle, depicted here, has largely gone.

The marshes are far from home-free, however. Now there are dams way upstream in the headwaters of the Tigris and Euphrates rivers in Turkey. That country's water-storage capacity on the Euphrates is now six times the river's annual flow. Its dams have halved average annual flows down the river into Iraq to 16 cubic kilometres. The dams have also suppressed the seasonal flood cycle so important to the health of the marshes. And Iran has not helped its wartime friends. It is diverting water from two of its own rivers than once topped up the marshes' waters, the Karun and the Karkeh.

The Marsh Arabs, once the guardians and managers of the Mesopotamian Marshes, have changed too, says Azzam Alwash, an Iraqi exile in California who set up an NGO called Nature Iraq to champion the restoration of the marshes.[160] 'They used to live isolated lives out among the reed beds. It was primitive and romantic,' he said. 'Today they mostly live in towns on the banks. They have cell phones and satellite dishes. They damage the fisheries with electric fishing. They shoot the birds. We see their garbage floating in the marshes.'

The marshes are unlikely to recover their former extent any time soon, Alwash says. And 'without a strong spring pulse of water down the rivers, the biology will change.' Even so, he is guardedly optimistic about the marshes' future. After all, the aims of the British in the 1950s and Saddam in the 1990s to obliterate them have been thwarted. And their resilience is now well established. He believes that the task for conservationists is to help the Ma'dan to fulfil the ambition they showed when tearing down Saddam's drainage structures two decades ago, to learn afresh in a modern setting how to live in harmony with the marshes. 'If the marshes were the Garden of Eden, then we continue to manage them like a garden.'[160] ○

RIVER RHINE, EUROPE
RIVER 'RECTIFICATION' REPLACED BY 'MAKING ROOM FOR THE RIVER'

WHAT IS A RIVER? TOO OFTEN, we still see rivers as little more than convenient conveyances for supplying water, removing waste, navigating ships, and getting rid of floodwaters. By those definitions, their wetlands and floodplains disrupt the 'proper' role of rivers – by slowing flows, consuming water, encouraging evaporation, and making the land beside them vulnerable to flooding. So, engineers have for centuries tried to improve nature's fluvial highways, by making them straighter, swifter, and more reliably contained within their banks. The notion that rivers need room to flood, or that they will defeat efforts to constrain them, has been an affront.

That certainly was the approach that for a long time governed management of the Rhine, Western Europe's longest river. It was crystallized in the early nineteenth century by German military engineer Johann Tulla. He was appointed by authorities along the river to redesign it according to the needs of the industrial age – a task he described as 'rectification'.[161] The word made clear that he and his sponsors felt nature had got things wrong, and the river needed to be made good.

For much of its course at that time, the Rhine went with the flow. It meandered across a wide floodplain of woodlands and water meadows. From Basel in Switzerland to Karlsruhe, then a fast-expanding German industrial city, a distance of around 200 kilometres, the river did not even have a fixed route. There was nothing to mark with a firm blue line on a map. It was braided, meaning it broke up into innumerable channels, which moved, disappeared, and reformed regularly.

This was inconvenient. Ships could not make it up- or downstream. Farmers were reluctant to cultivate the 100 square kilometres or more of seasonally flooded land each side of the river. Moreover, the border between Germany and France, which was supposed to follow the river for most of that distance, shifted whenever the channels moved. The hedonistic river defied the new expectations of what a river should be like. 'As a rule,' Tulla said, 'no stream or river, the Rhine included, needs more than one bed.' He set about forcing the Rhine into a single, permanent channel.[162]

From the day in 1812 when Tulla's first cut was made outside Karlsruhe, the project was controversial. Some villages complained that they would lose land to neighbours as the river's bed was shifted. Downstream in Prussia, then an independent state, they warned all too presciently that the rectification could worsen floods by constricting the river and rushing its water towards them, engulfing bankside cities such as Mainz and Mannheim.[163] Nonetheless, Tulla went ahead.

Tulla's blueprint took many years to achieve, and was only completed several decades after his death in 1828 from malaria, probably contracted while he was at work in riverside marshes. The rectification of the Rhine remains Germany's biggest ever construction project.[164] It delivered many of its short-term aims. Most of the backwaters, braided channels, and river islands disappeared. Industrial cities such as Cologne, Düsseldorf, and Koblenz were able to grow along the river's stabilized banks, and could be serviced by large ships. The river was kept in its allotted channel. Floodplains were reduced. The border with France was fixed. Malaria was eventually banished.

As time passed, riverside communities added their own piecemeal dykes, constricting the river further and turning more and more seasonally flooded pastures into arable fields. The upper Rhine was

The untamed River Rhine near Sargans in Switzerland, as seen by artist Johann Ludwig Bleuler in the early nineteenth century.

eventually cut off from 90 per cent of its former floodplain. The new, straight Rhine was 100 kilometres shorter than the old river.[165] But shorter, as the Prussians had predicted, also meant faster. The river now bypassed floodplains and meanders that had once diverted and slowed flood surges coming down from the Alps. Melting snow could now reach Karlsruhe in thirty hours, half the time it used to take.

Worse, peak floods in different tributaries of the main river now often coincided. As a result, flood surges that previously were likely only once every 200 years now occurred every sixty years. Flood peaks

in Cologne were a third higher than before. Tulla had been known in his time as the 'tamer of the wild Rhine'. But far from subduing the river, he and his successors had turned a slow, muddy stream into a river far less able to contain extreme floods.

The river's ecology was transformed too. Once, the Rhine and its floodplain functioned as a single ecosystem, extending from the Alps to the North Sea. Fish migrated up- and downstream and in and out of the river's floodplain and backwaters. But now the river was largely separated from its floodplain, and scraped bare of the gravel beds that provide fish with breeding grounds. In 1890, fishers caught 150,000

salmon on the Rhine. A century later, they caught fifteen. Smoked sturgeon was once a popular Dutch delicacy. But no sturgeon have been seen in the Rhine's waters since 1952.

The Rhine was far from alone in suffering the straitjacket of rectification. By the 1980s, only 30 per cent of West German floodplains remained connected to their rivers; the remaining 70 per cent were barricaded behind dykes.[166] But people were starting to have second thoughts about all this engineering. There was concern

Can we return the Rhine to its former glory? The image on the left shows the dyked Rhine near Sargans in Switzerland as it is now. On the right is an artist's impression of how the same spot might look with the dykes removed.

about the loss of wildlife along the Rhine, and a growing recognition that losing floodplains was worsening floods. A turning point came in 1995 when massive floods breached the river's banks for hundreds of kilometres. Downtown Cologne found itself under 2 metres of water. Politicians blamed Germany's then Environment Minister Angela Merkel. But far from accusing her of failing to raise dykes high enough, the complaint was that she had failed to order the flooding of low-lying meadows upstream to hold back the waters.

RIVER FLOODPLAINS 153

The same sea-change in thinking about floods was also taking place downstream in the Netherlands. There, the 1995 floods felt like an existential threat to a nation largely created by reclaiming land from river and sea. The country was for a while almost submerged. *Polders* – dyked areas of land designed to keep out floodwaters – were filling up like bathtubs. A quarter of a million Dutch people were evacuated from their homes. So were more than a million cattle, pigs, and sheep. Like their German counterparts, the Dutch responded with calls to make room for the river.

They dusted off a plan drawn up in 1991 that proposed achieving flood protection by returning 15 per cent of the country's drained farmland to soggy nature. The once-heretical call quickly became mainstream thinking.[167] As a trial, engineers breached a dyke at Duursche Waarden on the River Ijssel, the most northerly of four branches on the Rhine Delta. The breach allowed the river to take a detour into 120 hectares of marsh and willow forest. Since then, the Dutch have been making holes in dykes rather than plugging them up.

One of the biggest examples of 'making room' was the breaching of conventional flood defences at Noordwaard south of Dordrecht, on a branch of the Rhine known as the Nieuwe Merwede. Completed in 2015, the breach reconnects the waterway to 44 square kilometres of floodplain. The symbolism of the breach was not lost on the Dutch, who are well-versed in their nation's drainage history.

Noordwaard is famous for being the scene of the catastrophic St Elizabeth's Flood in 1421, when twenty villages were lost and as many as 10,000 people drowned in their beds. The flood inundated around 500 square kilometres. In places, the water soon retreated; but in Noordwaard, a giant inland sea formed. It was after that calamity that the Dutch got serious about dykes and drains. They became symbols of national aspiration. The Calvinist clergy lent God's support. 'The making of new land belongs to God alone, for He gives to some people the wit and strength to do it,' wrote Andries Vierlingh, a sixteenth-century Dutch water engineer.[168]

Over centuries, Noordwaard's farmers gradually reclaimed many of their fields, and eventually consolidated them into a single polder. But even their skills had limits. Some 80 square kilometres remained flooded and eventually became the Biesbosch National Park, a watery world occupied by beavers, white-tailed eagles, ospreys, kingfishers, and much else. To some this flooded land in the heart of a drained Dutch farming landscape denoted failure. But by the mid-twentieth century it began to inspire environmentalists. They wanted more

The Rhine in flood near Cologne.

such places. The Biesbosch park museum has a prominent picture of one of them, Wil Thijssen. He refused to repair a dyke that broke and flooded his land, and was condemned in the national parliament as a traitor. Now he is seen as a visionary, and his refusal as 'the first act of fluvial restoration in the Netherlands,' says his son Sylvo, who is Chief Executive of the state nature agency, Staatsbosbeheer.

In 2009, Thijssen's rebellion finally bore fruit. Noordwaard was 'depoldered' as part of the national drive to give rivers room.[169] Deprived of its dykes the former polder is now, in effect, an occasional overflow reservoir for the Nieuwe Merwede. It is flooded once or twice a year, whenever water in the river rises more than 2 metres. Local farmers don't like the scheme, says one of the engineers involved, Arno Schikker of construction company Boskalis. They feel their land has been sacrificed to lower water levels in the city of Gorinchem downstream.

Others certainly appreciate the new waterscape and its wildlife. On a weekday visit, we saw many Dutch birders offloading their bikes at Dordrecht Station and cycling along the lowered dykes, binoculars round their necks. But Dutch hydraulic engineering is never far away. Much of the country is still below the level of the rivers that flow through it. As we watched one group of cyclists pass, a giant barge loomed on the river behind. It was sailing several metres above their heads.

Ice forms on the inundated floodplain of the Lower Rhine in the Netherlands.

IN EUROPE, 1995 WAS THE YEAR when engineers realized that civilization could not carry on fighting the forces of nature in as blunt and full-on a way as before. Separating rivers from their floodplains almost always increases peak floods downstream.[170] However high you raise their banks, rivers in flood will find the weakest spot and burst through. Rivers should be allowed back onto their floodplains. Though with 10 per cent of Europeans reckoned to be living on river floodplains, that won't always be possible.

Following the Dutch model, Germany introduced legislation to 'give our rivers more room again; otherwise they will take it themselves', as federal Environment Minister Jürgen Trittin put it in 2003.[171] That meant reinstating some of the 1,200 square kilometres of Rhine floodplain lost since Tulla got to work. Dyked and drained fields were to be replaced by reed beds and water meadows that could flood every winter. The target was to shave 60 centimetres off flood peaks by 2020.

On the Elbe, the Danube, and other major European rivers, the ambition now is to tear down barriers, recreate old meanders, revegetate river banks, and above all to reconnect rivers with their long-lost floodplains. It will be a big task. The European Environment Agency's Director says his staff have counted over half a million human-constructed barriers on the continent's rivers, whether dams, weirs, or dykes. That is more than one for every 2 kilometres of river.[172] ○

SECTION THREE

INLAND SEAS, SWAMPS AND SUMPS

Lakes and inland seas are among the world's best-loved places – and among the most vulnerable. Without a regular supply of water they die, and may leave behind desert, like the Aral Sea in Central Asia. But their magic can also be a focus for wider conservation and cross-border peace-making, as has happened at Lake Prespa in the Balkans. And we can create new ones, either deliberately or accidentally. In the Anthropocene, drainage sumps such as the Salton Sea in California have become vital homes for wetland wildlife.

PREVIOUS PAGES: Deprived of its river inflows, the Aral Sea has been drying up for half a century. Here in Uzbekistan, the shore has retreated by more than 100 kilometres.

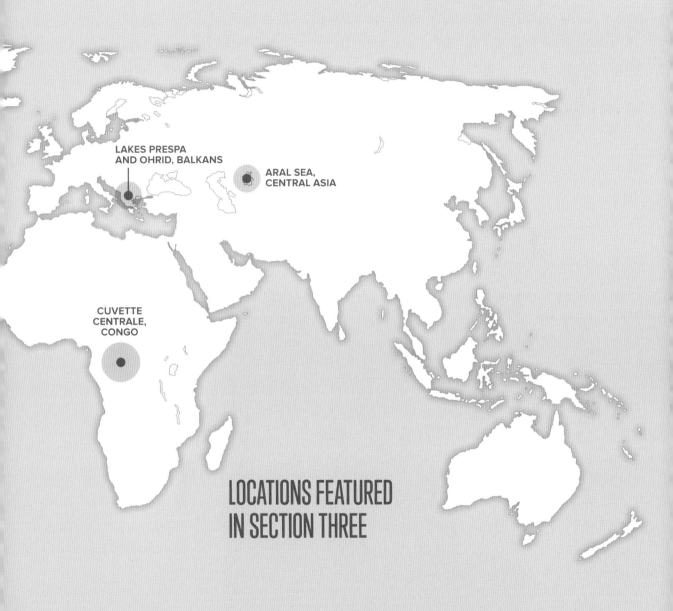

INLAND SEAS, SWAMPS AND SUMPS 161

SALTON SEA, CALIFORNIA
CELEBRATING ARTIFICIAL AND ACCIDENTAL WETLANDS

THE SALTON SEA IS A DRAINAGE SUMP in the desert of southern California – an artificial wetland, whose contents are the polluted outpourings of farms in the Imperial Valley. But in a state where 90 per cent of natural wetlands have been drained, it has become California's best bird habitat, home to 380 species. The sea shouldn't be there. It came about by accident a century ago, when land speculators decided to turn a sweltering desert depression known locally as the 'valley of death' into irrigated farmland. Charles Rockwood and George Chaffey went about their task in the most cavalier manner imaginable, by capturing some of the flow of the mighty Colorado River, 100 kilometres to the east in Yuma, Arizona, and diverting it down a rickety canal. The project soon went spectacularly wrong when the river began to pour its entire contents down the canal. The fields at the other end swiftly became a vast and rapidly growing inland sea.[173]

The government was forced to intervene. It took the contents of 6,000 railroad cars loaded with stones and gravel to restore the Colorado to its rightful course. Afterwards, the farmers who had been first flooded out and then left high and dry, successfully lobbied the government to give them a new permanent canal to irrigate their fields. The All-American Canal was eventually completed in 1938. Today, the Imperial Irrigation District is the largest single recipient of water from the Colorado.[174] Its farmers have grown rich. And a smaller version of the inland sea lives on at the valley's lowest point, receiving the salty, chemicals-laden run-off from their fields.

Despite such unpromising beginnings, it didn't take long for the Salton Sea to develop its own ecosystem. Some species like heavy doses of fertilizer, salt, high temperatures, and sometimes low concentrations of oxygen. A rich soup of plankton and algae fed fish such as African tilapia, which were introduced in the 1960s. The Salton Sea was soon one of the world's most economically productive lakes.[175] Birdlife prospered too, as the fish attracted egrets, cormorants, brown pelicans, and various boobies.

The weird ecology has been accompanied by a weird cultural cachet. In the 1960s, before Las Vegas emerged as a desert oasis for the jet set, the Salton Sea was the place to be seen. Frank Sinatra sang there. With him came yacht clubs, beauty contests, and nightclubs. Even the Beach Boys put in an appearance on a shorefront that had everything except surf.[176]

But the artificial desert oasis has been living on borrowed time and borrowed water. There has been growing competition in California for the Colorado's water. Nearby San Diego, a city of more than a million people, is growing thirsty. It gets priority over Imperial Valley farmers, who are now being incentivized to use their water more efficiently, and to pass on the surplus. But less wasted water means less water draining into the Salton Sea. 'On-farm conservation' is causing 'off-farm Armageddon', as a former Director of the Salton Sea Authority, Tom Kirk, warned.[177]

On current expectations, the sea's volume will probably halve by 2030, exposing more than 200 square kilometres of the lake's bed, says Michael Cohen of the Pacific Institute, a California water think-tank.[178] The accidental wetland will never be the same again. But with careful management, a much smaller Salton Sea could remain an important oasis in the desert, with birds inhabiting carefully managed small wetlands around its fringes.

THE WORLD HAS MORE AND MORE such places. We should be grateful that accidental, artificial, and leftover wetlands are growing in extent faster than natural wetlands are retreating. Recent satellite analyses

The Salton Sea in southern California is an artificial wetland in the desert, collecting drainage water from irrigated farms. Recent action to save water by making irrigation for agriculture more efficient means there is less drainage and the sea is shrinking.

PAGE 162:
Burrowing owls guard their nest as the sun sets over the Salton Sea in California. But if the sea dies, where will they hunt?

show that new permanent water surfaces around the world cover 184,000 square kilometres, with another 29,000 square kilometres achieving a seasonal covering.[179] Some have been created deliberately, for recreation and conservation, to revive urban wastelands or improve flood protection. More often they are useful by-products of other land uses, or are entirely accidental. However they happen, as natural wetlands are drained, dyked and dammed, their importance for wildlife can only grow. They are the new reality – and a new resource – for the Anthropocene.

One case is the 125 square kilometres of former salt marsh in the eastern suburbs of the Indian megacity of Kolkata. For many decades, engineers poured cleaned-up effluent from the city's main sewage works into the marshes. When the sewage works ceased to function, many expected the marshes to clog with raw sewage and become biologically dead. Instead, the sewage pushed the marsh's ecosystems into overdrive. The story was first told by Dhrubajyoti Ghosh, a city sanitary engineer.[180] He found that the wetlands' rich colonies of algae were happily gobbling up the sewage bacteria. In effect, they had replaced the sewage works as a cleansing agent. And, with more

sewage in the lake, there were more algae. Fish such as carp and tilapia grew fat on the abundant nutrients. Fishers were delighted.

Decades on, the natural sewage works is alive and well. It processes some 750 million litres of sewage a day, and feeds enough fish to sustain annual catches of thousands of tonnes. The fishing provides livelihoods for an estimated 30,000 poor people of Kolkata. More than that, farmers syphon off the nutrient-rich waters to irrigate and fertilize rice fields and vegetable gardens along the wetland's banks. Some may worry about the health consequences, but half the vegetables in the city's markets come from these fields. Nature does well, too. The East Kolkata Wetland is home to several species of mongoose, forty species of birds and a plethora of snakes. In 2002, the world's largest organic sewage treatment system was declared a wetland of international importance by the Ramsar Convention on Wetlands.

The most widespread artificial wetlands are reservoirs that form behind dams. Many reservoirs have been thought good enough to gain conservation recognition. Dozens, from Rutland Water in England to the Kayrakkum Reservoir in Tajikistan, are now listed by the Ramsar Convention. They may sometimes partially compensate for wetlands downstream damaged by the dam's interference with river flow. In Tunisia, for instance, an estimated 190 square kilometres of natural wetlands have been lost in the past century, replaced by 220 square kilometres of artificial reservoirs.[181]

But reservoirs rarely mimic natural wetlands perfectly. Wildlife finds it hard to adapt to water levels that go up and down with the demands of customers for water or hydroelectricity. Also, the reservoirs are often in mountainous terrain, with steep banks that give little opportunity for the development of shoreline marshes. The River Guadalquivir in Spain has thirty dams in the mountains, whose reservoirs have become important wintering and breeding habitat for ducks, coots, and reed warblers.[182] But that is hardly recompense for the loss near the river's mouth of four-fifths of the Doñana Wetland, one of Europe's most biodiverse ecosystems with more than 4,000 species.[183]

Nevertheless, there are ecological riches to be had. Take industrial salt pans, known as *salinas*. These highly engineered networks of interlinked water basins are often established on natural coastal lagoons. They are designed to evaporate sea water and manufacture salt. But they are often also extremely valuable wetland habitats.

The largest salina in Europe is the Salin de Giraud. It covers 110 square kilometres of the Camargue Wetland in the largest delta in the European Mediterranean, at the mouth of the River Rhône in

PREVIOUS PAGES:
Dead palm trees stand on the former shoreline of the Salton Sea, a reminder of the time when it was a fashionable resort.

southern France. It has more than 100 basins, with concentrations of salt that range from that of the seawater pumped into the lower lagoons to levels ten times higher. In the lower lagoons salt-tolerant plants and fish thrive, but at higher concentrations brine shrimp are about the only aquatic inhabitants. Visiting birds can pick and choose the salinity and food supply of their choice.

The Camargue as a whole is a maze of sand bars and reed-covered marshes grazed by its famous breed of white horses and bulls, which are exported to Spain's bull rings. But the brine ponds of the salina hold arguably the biggest ecological prize, more than 10,000 pairs of pink-plumed greater flamingos. The birds breed on an artificial breeding island with nesting mounds created by conservationists from the nearby Tour du Valat research station.[180] If the pumping of saline water between the Salin de Giraud's lagoons was not maintained, the birds would be in dire straits. This is their largest population, and the only other breeding habitat in the region is also on industrial salt pans, in Malaga and on the Ebro Delta.[180]

With such riches, no wonder abandoned salinas are often taken over by conservationists to maintain their saline cascades as bird habitat. Thanks to their work, Malta's Ghadira Wetland reserve, for instance, is still visited by 140 species of migrating birds, such as redshanks, sandpipers, and egrets.

Agriculture also provides inadvertent artificial wetlands. From the deltas of the Mediterranean to former mangrove swamps across Asia, irrigated rice fields are often rich in water plants and fish. All told, they cover an estimated 1.6 million square kilometres, six times the size of the UK.[184] Their ecology is under-researched, compared to more 'natural' wetland ecosystems. But where there are fish, there are usually water birds stopping off during migrations, foraging for food, or nesting in the crops and fringing vegetation.[185] On the delta of the River Ebro, Spain's second longest river, coots, mallards, shovelers, and widgeons move onto rice paddies after the autumn harvest. They are joined over the winter by thousands of waders. Japanese researchers reckon that half of that country's endangered plants and animals live in or near rice paddy.[186]

Farmers also dig ponds to provide water for livestock. British countryside surveys for many decades showed a decline in such ponds. But in the decade from 1998 to 2007 their numbers increased by 50,000 to almost half a million. Ponds can support two-thirds of Britain's freshwater species.[187] New ponds are better than old ponds, the surveys found.[188] Britain is far from alone. 'Ponds [both natural and artificial] are the most common and widespread habitat for all plants and animals across the continents,' says Mike Jeffries of Northumbria University in England.[189]

Past mining developments and flooded former industrial sites can pass muster as valuable wetlands, too. Europe's largest population of great crested newts – about 30,000 of them – lives happily in ponds created for them in old brick pits outside Peterborough in eastern England. Many British holidaymakers on the waterways of the Norfolk Broads in eastern England have no idea that they are sailing on flooded medieval peat excavations. And the Thames Estuary has

LEFT: Marker Wadden in the Netherlands – a new artificial island created to enrich the ecology of an artificial lake.

RIGHT: Dawn mist evaporates from the Norfolk Broads in eastern England, revealing the shape of a drainage mill. The waterways of the Broads are the flooded legacy of medieval peat workings.

a series of lagoons and salt marshes on former industrial sites that have become rich in wildlife. West Thurrock Lagoon, for instance, is on the site of a demolished riverside power station. It has more rare species than almost any other nature site in Britain, including thirty-six different types of bees. It is one of only two known UK sites for the distinguished jumping spider.[190]

ONE OF THE MOST AMBITIOUS deliberate artificial wetlands is the Marker Wadden in the Netherlands. It is an attempt to make a virtue out of a mess created in 1975, when engineers cut off a portion of former sea known as the Ijssel Meer with a dam. The original aim had been to drain the *meer* and turn it into farmland. When the Dutch lost interest in doing that, their Plan B was to make a water park for swimming, sailing and birdwatching. But the meer began accumulating sand and silt brought in by the River Ijssel. The inputs

smothered bottom-dwelling organisms and turned the waters into a turbid soup that caused fish and bird populations to crash. In short, the meer was dying. So, coastal engineers came up with a clever Plan C. They proposed dredging up the mess to make an island. The government jumped at the chance. Within a couple of years, the dredgers turned 13 million cubic metres of sand into a shoreline that enclosed a similar volume of silt, poured into the middle to make a marshy interior.

A twenty-minute boat ride from the small port of Lelystad, the 800-hectare island is an impressive and attractive site, with sand dunes, mussel beds, willow stands, and reed beds. It is billed as one of the largest nature restoration projects ever attempted in Europe. Birds have started to show up, starting with terns and black-headed gulls. In late 2018, tourists began making day trips from nearby Amsterdam, to tread the island's boardwalks.

It feels as much theme park as nature restoration. The test will be whether, once complete, it can maintain itself without constant replenishment of the materials that made it. Its designers hope the island will act as a silt trap in the meer, cleaning it up by naturally capturing the remaining silt.[191] But things may not be so simple. Rather than growing, the island may wash away, dispersing the silt. When we visited in mid-2018, engineers were doing running repairs after ice from the meer had eroded the foreshore the previous winter. We shall see. But that is likely to be the nature of artificial and accidental wetlands. And sometimes we will be pleasantly surprised at what nature can make of our messes. ○

LAKES PRESPA AND OHRID, BALKANS
WETLAND TRUCE BRINGS WIDER PEACE

THE INSPIRATION FOR HIS LIFE'S WORK began for Rogos Catsadorákis one day as a teenager in the 1970s, when he saw a TV programme about the Dalmatian pelicans of Prespa, a large lake in the heart of the Balkans. He was entranced. The programme told how 'for decades, ornithologists had searched for where Dalmatian pelicans nested in south-east Europe,' the Greek conservationist says today. Then they found it. The home of the world's largest freshwater bird was in Lake Prespa, right on the border between Greece and Albania.

After graduating, Catsadorákis took the first opportunity to leave his home in Athens and head for the lake. 'I was almost the first outsider to stay for any length of time,' he says. And he has devoted much of his life to securing the pelican's future – in a region of Europe where the safety of humans could for a long time not be assured, let alone the survival of the pelicans. His story became the story of the lake itself, a good-news story of how cross-border environmentalism can help bring about wider peace.[192]

By the early 1980s, Greece had joined the European Union. Brussels was throwing money at the area around the lake, which was depopulated and recovering from a series of international conflicts. The Greek Government was requesting EU funds to develop farming on the wet meadows around the lake in ways that would disturb the birds' breeding grounds. 'Back then, they had no thoughts about conservation,' Catsadorákis says. Most people living near the lake had long regarded the pelicans as vermin. 'There was a state of war between fishermen and the pelicans. Locals would kill the birds and take their heads to police stations and receive a cash reward,' he says. 'But I was bewitched by them and the mystery surrounding them. I wanted to understand their lives and how to protect them.'

So, he and a growing band of ornithologists set about winning hearts and minds, slowly persuading fishers and others who went to Mikri Prespa, a small sub-lake where most of the pelicans breed, to keep away from the colony. 'At the start, they were very suspicious of us, and said it was their right to fish there. But little by little they realized there was nothing sinister about us.'

The birders set up the Society for the Protection of Prespa. It built artificial nesting rafts to float on the lake, and in 2006, amid growing concern that Mikri Prespa was drying up, constructed a sluice gate to hold water inside. The gate's operation encapsulates the new accord. It opens and shuts according to a schedule agreed between conservationists and local landowners, fishers, and other stakeholders. It maintains good breeding conditions, while avoiding floods near the lake shore that would prevent cultivators from cutting hay, or herders from grazing their cattle.

Almost four decades after Catsadorákis first arrived on the lake's shores, he has much to be proud of. 'Now, the locals are proud of the birds that they once enjoyed killing,' he says. 'From dwindling numbers, we have created a strong source population of pelicans for the whole region.' But his fight to protect the birds remains work in progress.[191] The habitat that sustains them is a product of informal agreements that must be maintained. 'We can't let things develop naturally now,' he says. 'If we did, Mikri Prespa would disappear, and with it the pelicans.'

Much the same could be said of the wider lake.

Greece is one of three nations with a share in Lake Prespa. From early on, Catsadorákis and his colleagues sought links across borders to protect the lake. Finding friends was not easy to start with. The

Just forty years ago, most people living around Lake Prespa regarded the pelicans as vermin. They received cash rewards for killing them.

lake had long been a crucible of conflict. In the late 1940s, there had been a civil war in Greece. Much of it was fought around the lake. 'Villages emptied and foreigners stayed away,' says Catsadorákis. Later, Yugoslavia and Albania feuded, and the lake became a no-man's land between them. But like other depopulated war zones, the lake for a time provided space for wildlife to prosper. And when ornithologists found that it contained the elusive breeding sites of Dalmatian pelicans, conservation became the key that unlocked wider peace-making between the three nations.

The first to turn that key was Greece's Foreign Affairs Minister at the turn of the millennium, George Papandreou. He negotiated a trans-border Prespa Park with Albania. That led to the signing of an agreement involving all three nations and the EU on conserving the lake. The new accord didn't stop there. In 2019, Greece and the former Yugoslav Republic of Macedonia resolved a long-running dispute over the latter's name. They called it the Prespa Agreement, and it was signed at a lake-shore hotel near Yugoslav strongman Marshal Tito's old country retreat, in what is now North Macedonia.[193] The lake has become a symbol of conciliation between the countries that share it.

BUT WHAT, EXACTLY, DO THEY SHARE? Words such as pristine, pastoral, and isolated are widely used to describe Lake Prespa. They have some meaning. The lake is certainly picturesque, extensively fringed with thick beds of reeds. It is also an important hotspot for nature. Of the eleven native fish species living in the lake, nine are found nowhere else.[194] In the mountains that surround the lake, there are still wolves and bears. But a tour around the lake, taking in all three countries, suggested a landscape and ecology more managed than pristine. And a lake where fine words about conserving its ecology need to be backed up by firm planning now that the no-man's land is becoming a valuable resource.

One of the lake's more impressive natural features is the Ezerani Wetland, a Ramsar-listed expanse of 20 square kilometres of reeds, wet meadows and open water on the northern shore in North Macedonia. At various times, say ornithologists, more than 40 per cent of the bird species known in Europe have been spotted here. But ecologists point out that the preponderance of reed beds here and elsewhere around the lake is not a sign of pristine nature, but a

Mikri Prespa, the southern portion of Lake Prespa. It took a decades-long search by Europe's ornithologists to discover that this is the main nesting area for Dalmatian pelicans.

symptom of too many nutrients reaching the water from the land. And while the reeds do help cleanse the lake of the nutrients, and provide nesting sites for birds, they also prevent fish from reaching spawning areas in the marshes and wet meadows beyond.

The marshes and meadows are also under siege in places from house building, speculative hotel construction and, especially, irrigated agriculture. In North Macedonia, the crop of choice is apples. At the Stenje Marsh on the lake's western shore, we walked towards a watchtower built to allow visitors to view its birds. But all around it were orchards. A few metres away, we stumbled on an illicit suction pump taking water from what remained of the marsh to new fields. 'This was a protected area, full of plants, insects, and birds,' said Daniela Zaec of the Macedonian Ecological Society. 'We used to bring students here. But every year, the farmers encroach further. Now all you can see from the tower is apples.'

On the Greek side, much of the lake shore is taken to irrigate beans. In Albania, the latest lakeside crop is *Sideritis*, a herbal tea mostly exported to Germany. Almost every lakeside village we saw was annexing grazing meadows to grow the crop. Despite such incursions, local environmentalists are guardedly optimistic about the lake's prospects. They say that things are often a lot better than in the reckless days after the fall of Communist rule. Aleksander Trajce, who runs Protection and Preservation of Natural Environment in Albania, spent much of his childhood on the shores of Lake Prespa. 'As a kid in the 1990s, I remember a lot of illegal fishing. It's much better now,' he said as we breakfasted in a small hotel on the Albanian shore.

Albanian Prespa remains poorer than its neighbours. There are more donkeys than cars; people cultivate or graze every available plot; and a good many communities still rely on private wells for their drinking water. But there is a sense of environmental order, too. The lake shore and the forested hills behind are in the hands of park authorities. Long-standing park manager Vasil Jankulla told us he had successfully cracked down on illegal logging in the forests near the lake, while still allowing locals to collect firewood for the winter. 'A few years ago, you could see the border between us and North Macedonia from far away, because it was forested on the Macedonian side but bare earth this side,' he said as we drove to the border. 'Now you can't see the difference. Our trees were as good as theirs.'

Jankulla wanted to use the forests to relieve pressure on the lake's meadows. His plans included a scheme to revive traditional ponds in the forests in the hills. The idea was that cattle could spend more time away from the shore. The ponds would attract wildlife too, he said, showing us video footage of a bear using one of the ponds.

The same pragmatic management applied to Albania's lake fisheries. The local fishers' association had been granted a ten-year concession to manage the lake fish stocks and police its own members. Trajce took us to Tuminec, a traditional Albanian fishing village, to see how it worked. The answer was precise rules on who could fish where, a proper closed season while fish were breeding, and a strong sense of collective ownership.

Over lunch in a lakeside tavern, we met Moza, a fisher woman in what is still largely a man's world. Cheery, strong, and red-faced, she bounced into the room, fresh off a bus back from the market in Korça, where she had been selling the fish she had netted the night before just a few hundred metres from where we ate. Her village's fishing territory stretches about 5 kilometres to the North Macedonian border, she said. Every fisher has his or her spot to catch indigenous fish such as the local carp, chub, minnow, trout, bleak, or nase.

She couldn't stay to eat, she said, because it would soon be time to get back in her boat to set her nets once more. She would collect the fish at midnight and be on the 6.30 a.m. bus back to market. We couldn't figure out when she slept. But she assured us fish stocks are good. 'The number of big fish are increasing. Smaller fish are only at the same level as before. But that is because some are being eaten by the pelicans.' Somehow, every story on Prespa comes back to the pelicans. Hopefully, said Catsadorákis, while they prosper everything will be well.

LAKE PRESPA IS ALL BUT AN INLAND SEA. It has only one outlet: an underground channel from the North Macedonian shore, through the limestone rocks near the village of Gorica e Vogël. You can stand on the road above and watch the water disappear. It comes out 15 kilometres away at the Saint Naum springs in the neighbouring Lake Ohrid.

For a long time, Lake Ohrid, which is divided between what is now North Macedonia and Albania, was more famous than Prespa. It was a scenic stopping-off point for northern Europeans journeying to classical sites in Greece. Edward Lear sketched the lake and the Studenchishte Marsh, a once-extensive peat swamp on its eastern

These Albanian forests on the hills above Lake Prespa have benefited from strict management that halts illegal logging while allowing seasonal cutting of firewood by local villagers.

The local trout are an important catch for fishers on Lake Prespa.

shore. Ohrid gained scientific fame in the twentieth century thanks to another outsider, Serbian biologist Sinisha Stankovic. Rather like Catsadorákis and Prespa, the lake's story wouldn't be the same without him. Stankovic devoted his life to exploring its species, setting up a research institute beside the Studenchishte Marsh in 1935 to complete the task.

The lake, Europe's oldest, turns out also to be of huge ecological importance. Stankovic catalogued more than 200 endemic lake plants, invertebrates and fish, including species with a habitat of only a few square metres, living in tiny ecological niches around the underwater springs where Prespa water bubbles into the lake. His book, *The Balkan Lake Ohrid and its Living World*, cemented the view that the lake is for its size the most biodiverse on Earth. Few disagree. Christian Albrecht of Liebig University in Giessen, Germany, calls Ohrid 'one of the few extraordinary lakes in the world ... a Holy Grail for biologists from all over the world'.[195]

Stankovic's institute is rather run down for lack of funds, these days. We took a tour with one of the great man's successors, fish biologist Trajce Talevski. In the basement, he showed us giant tanks still in use to nurture the endangered Ohrid trout. They produce around a million eggs a year that help sustain the lake's stocks. But such efforts to sustain the lake's unique ecology have been threatened by plans to develop the historic city of Ohrid, on the lake's northern shore, into a hotspot for mass tourism, complete with cheap flights from London's Luton Airport throughout the summer.

Hotel developers are demanding every scrap of land they can obtain, said Vladimir Trajanovski from SOS Ohrid, a citizens' initiative dedicated to protecting the lake. Touring Ohrid city, he showed us public beaches concreted over, restaurants extended on stilts over the lake's clear blue waters, and secluded coves where we watched restaurant workers dump their garbage. But most worrying for the future of the lake's ecology was the spread of the town east onto the Studenchishte Marsh.

The marsh was once the lake's most important area for fish spawning, and remains a vital hotspot of biodiversity in its own right, known for amphibians such as Macedonian crested newts and yellow-bellied toads.[196] Once it stretched for miles round the lake shore, but today just 70 hectares of its reed beds and pools remain. They are largely cut off from the lake by cafés and a roadway funded by the European Union. Only a vociferous campaign in 2015 by biologists from across Europe persuaded the city's mayor to back off from a plan to drain the remaining marsh.[197] 'The death of the marsh would be the death of the lake,' said Nadezda Apostolova, of the University of Valencia.

It could be just a brief stay of execution, however. A walk around the marsh showed it hemmed in by a landfill, a tree nursery, an army post, fly-tipped construction waste, and impromptu houses. There were plans for a 44-hectare tourist complex at Gorica just south of the marsh. But there is at least a chance that the development juggernaut may have been halted, said Talevski. Startled by the pushback from foreign scientists,[198] some in the city's tourist trade would like a greener image. Conservationists have come up with plans to create a tourist attraction by rewetting some of the lost marsh and reconnecting it to the surviving reed beds.[199]

THAT COULD BE A GOOD SIGN for the rest of the Balkans, too. The region has been facing a tsunami of hydro-engineering projects in recent years, with dozens of land drainage, river straightening, and hydroelectric dam schemes proposed. Some have been earmarked for largely wild rivers, such as the Vjosa in Albania, with potentially devastating consequences for river and wetland ecosystems all the way to the sea.[200] Researchers estimate that a tenth of Europe's fish species could be at risk. But as the nations of the Balkans rush to catch up with economies in Western Europe, they seem to have missed an important lesson learned the hard way there – about the folly of indiscriminate concrete pouring.

Balkan environmentalists are drawing on the lessons about cross-border cooperation learned on the shores of Prespa, however. They are fighting back. A big cheer went up across the region in the final days of 2018, when a group of protesters known as the 'brave women of Kruscica', who had for 500 days occupied a bridge leading to a dam construction site in Bosnia and Herzegovina, were able to go home for Christmas.[201] They had successfully persuaded judges in Sarajevo to prevent construction of two dams on the River Kruscica. Across the Balkans, environmentalists are securing important victories in the face of reckless development proposals of a kind now rarely seen elsewhere on the continent. The spirit of Prespa lives on. ○

ARAL SEA, CENTRAL ASIA
WHAT HAPPENS WHEN A SEA DIES

A LITTLE OVER HALF A CENTURY AGO, the Aral Sea in Central Asia was the world's fourth largest inland sea. It covered an area the size of the Netherlands and Belgium combined. It was renowned for its blue waters, plentiful sturgeon, stunning sandy beaches and bustling fishing ports. Hundreds of trawlers set sail from Muynak in Uzbekistan on its southern shore to harvest the sea's fish, which were sold the length of the Soviet Union. Kremlin top brass spent their holidays there. But the same leaders also decreed that the waters of the two rivers that fed the sea should be captured to irrigate cotton fields across the region. The Amu Darya, once known as the Oxus, had once been bigger than the Nile, but soon delivered barely a trickle to the sea most years. The Syr Darya managed less than a quarter of its former flow.

The diversions have continued in post-Soviet times. With little new water, the Aral Sea has been left to evaporate in the sun. It survives today as two hypersaline pools, containing a tenth as much water as before, plus a fishing lake in the Syr Darya Delta, misleadingly named the Small Aral Sea.[202,203] Stand on the old beach promenade at Muynak and the sea is more than 120 kilometres away across virgin desert. The town's old port is full of rusting trawlers. A few miles away, in the fly-blown coastal town of Uchsai, which once had a fish-smoking plant, a teacher remembered, 'The sea left here in 1961. It came back in 1966 and 1968, but that was the last time we saw it.' The UN-backed International Fund for Saving the Aral Sea, whose launch one of us

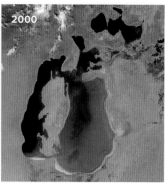
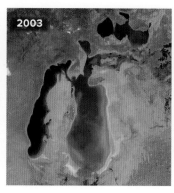
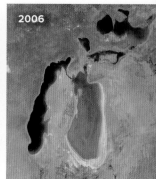

(FP) attended in the delta town of Nukus in Uzbekistan in 1992, has done little to address the calamity. Its 2018 meeting was the first in nine years.

The loss of the sea's large surface of water has made the climate across Central Asia more extreme, with intense heat as well as cold. The old people of Muynak remember how they used to swim in the sea as late as November. Now, they wear overcoats in the town in October. The wide expanse of the Amu Darya Delta, where the Caspian tiger once hunted, is desert. The tiger is extinct. Dust storms whipped up by winds howling across the exposed sea bed carry salt and toxic agricultural chemicals that have created a health crisis in former coastal communities. Since the sea died, life expectancy has plunged and infant mortality soared.[204] Anaemia has become endemic. Many people have left. Muynak's population has fallen by three-quarters to just 10,000. The drying of the Aral Sea is one of the great environmental tragedies of the twentieth century – a stark lesson about what happens when water is grabbed from rivers without limit.

AND SO IT GOES. Landsat data show that the world has around 117 million lakes.[205] Most are small bodies of water in wider wetlands, many of them seasonal. But the world's permanent lakes have lost some 90,000 square kilometres of surface area since 1984, with another 62,000 square kilometres changed from permanent to seasonal lakes. Most of the loss has been in just five countries, all in arid Western and Central Asia. They are headed by Kazakhstan and Uzbekistan, which share the dried-up shores of the Aral Sea.

The other big losers in the Landsat analysis are Iraq, home of the Mesopotamian Marshes, Iran, and Afghanistan. On the border between the latter two lies the 5,700-square-kilometre Hamoun Wetland. Until recently, Hamoun's three lakes and surrounding

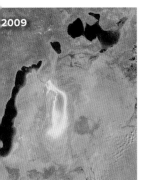
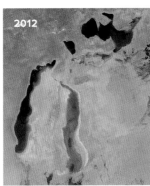
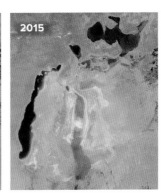
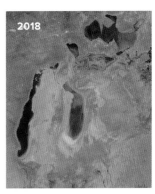

The death of the Aral Sea in Central Asia. This time series of satellite images shows its demise from 1989 to 2018, as the rivers that once kept it full were diverted to irrigate cotton.

FOLLOWING PAGES:
The rusting remains of beached trawlers at Muynak, once a port on the southern shore of the Aral Sea in Uzbekistan. The inland sea, once the fourth largest in the world, began receding after Soviet irrigators took its waters to grow cotton. The shore of the sea's surviving remnant is now more than 100 kilometres to the north.

marshes formed an inland delta of the Helmand River, flowing out of Afghanistan. The lakes were home to 140 species of fish, and an important staging post for migrating birds. Leopards hunted and otters swam. Hamoun may be less famous than the Mesopotamian Marshes, but it has a human history at least as illustrious. It was for thousands of years one of a handful of reliable watering places in the region, a vital source of food, shelter, and refuge along the old Silk Road. It prospered under numerous empires and several religions.[206] It is where Zoroastrianism originated.

But in the past twenty-five years, a combination of an Afghan dam on the River Helmand and Iran's own water diversions has rendered the wetland all but lifeless. The river is choked with tamarisk. The lakes are dry. The reed beds have become salt flats. Villages lie abandoned.[207] Most of the quarter of a million Sistani people who once lived there have retreated to Zabol, the area's only city, or live in refugee camps.

The wetlands of this part of the world have numerous such casualties. The Dead Sea, famously the saltiest lake on the planet, has for decades been deprived of water from the River Jordan. Israel takes most of the river's water, with Syria and Jordan consuming the rest. Since 2008, the Dead Sea has divided into two parts. Its shore is in

INLAND SEAS, SWAMPS AND SUMPS 185

places quicksand. Sinkholes swallow cars. An area around the kibbutz of Engedi has been abandoned.[208]

A hundred kilometres to the east, the Azraq oasis has been populated longer than almost anywhere in the Middle East. For millennia, caravans stopped there while crossing the desert from Arabia to the Mediterranean. The oasis is the collecting point for rain falling across a basin larger than Belgium. But since 1980, most of the water reaching the oasis has been pumped out to fill taps in the fast-growing Jordanian capital of Amman. Despite a UN-backed emergency rescue plan,[209] the oasis is still only around 4 per cent of its former size.[210] Just enough to build a pleasing boardwalk for tourists.

Water abstraction is much the most frequent cause of the demise of lakes and inland seas. But it is not the only risk. In some places, physical damage to the bed of a lake creates havoc. Take China's largest freshwater lake, Lake Poyang on the floodplain of the Yangtze River, where 90 per cent of the world's Siberian cranes spend the winter. Since 2000, it has become the largest sand mine in the world, supplying a construction boom down the river in Shanghai. Dredgers have removed more than 400 million tonnes annually. They have doubled the size of the waterway that links the lake to the Yangtze, which has partially drained the lake. The cranes have survived so far, but there has been a large decline in lake fisheries and a collapse in the population of the river's critically endangered finless porpoise.[211]

In almost all cases, lakes and inland seas can recover. Only a few years ago, Lake Urmia in north-west Iran seemed doomed to disappear.[212] Between 2000 and 2015, dams and upstream irrigators deprived the world's second largest salt lake of 90 per cent of its water. Its famous flamingos abandoned breeding islands because their favourite food, a brine shrimp, had gone.[213] But pragmatic action is turning things round. Iran's president Hassan Rouhani announced a lake restoration programme. His government ordered water releases from the dams and encouraged farms to use water-efficient drip irrigation and plant drought-tolerant crops. A startling revival occurred,[214] helped by heavy rains in 2019. According to the UN Development Programme, which helped, half the lake has recovered. If the Aral Sea is a totem of reckless mismanagement of water on a continental scale, then Lake Urmia shows that, even in a desert, the water can be brought back. ○

CUVETTE CENTRALE, CONGO
JUNGLE SWAMPS AND A CLIMATE TIME BOMB

THE WATERLOGGED JUNGLES of the Congo Basin are, for Europeans at least, both dream and nightmare. The Victorian explorer and journalist Henry Morton Stanley, who first described Africa as 'the dark continent', required 356 porters to take him through. Many of them drowned. In scarcely less cavalier a manner, the contemporary Irish explorer Redmond O'Hanlon crossed the swamps in search of Lake Tele and a mythical Congo dinosaur, a story told in his book, *Congo Journey*.[215]

Then there was Simon Lewis, an earringed young geographer with some of the eccentricity of his forebears. With colleagues from Leeds University in England, he followed a nearly identical route to O'Hanlon, crossing the Cuvette Centrale, a swamp full of gorillas and elephants – if not dinosaurs – on the west bank of the River Congo. Lewis took a peat auger with him, a simple tool to drill into the swamp to discover the contents of its soggy soils. In so doing, he discovered the world's largest deposit of tropical peat, up to 6 metres thick and covering an area larger than England. The peat contains an estimated 30 billion tonnes of carbon, accumulated over 10,000 years.[216] That, Lewis told an astonished scientific world after exiting the swamp in 2017, is as much carbon as in all the trees in the rainforests of the Congo Basin, and the equivalent of about three years of global fossil fuel emissions.

The discovery came just two years after another team of British geographers had emerged from the jungles of north-east Peru to report another previously unmapped tropical peatland, covering 35,000 square kilometres of the upper Amazon Basin, south-west of the jungle city of Iquitos. Above ground, the Pastaza-Marañón swamp is covered in palm trees and inhabited by tapirs, monkeys and macaws.[217] Below ground lies 'the most carbon-dense landscape in Amazonia,' according to Katy Roucoux of the University of St Andrews.[218]

The two finds, combined with earlier discoveries of rich peat deposits in waterlogged swamps on the Indonesian islands of Sumatra and Borneo, have transformed the world's understanding of peatlands. Previously, scientists had assumed that the planet's peat – which is essentially the remains of undecayed vegetation – was almost all in cold climates, such as Siberia, the Canadian Arctic, Alaska, and northern Europe. 'It seemed counterintuitive that peat could accumulate in hot tropical environments where plant litter decomposes very fast,' says Ian Lawson, a member of the St Andrews team working in Peru. 'But now we realize that waterlogging is more effective than had been assumed in slowing the rate of decomposition, regardless of temperature.'

That means there could be a great deal more undiscovered tropical peat. 'There are lots of floodplains in Amazonia and in other parts of the tropics that are yet to be looked at,' says Lawson. 'There is still no substitute for going out there with a peat auger.' But while waiting for intrepid explorers with wading boots and augers, Thomas Gumbricht of the Center for International Forestry Research in Bogor, Indonesia, has used Wetlands International emissions data[219] to carry out a modelling study of the geological and climatic conditions that would be likely to create peat bogs.[220] He reckons peat may lie beneath 1.7 million square kilometres of the tropics, four times more than currently mapped. He suggests explorers take their augers to the vast flooded forests along the Amazon and Orinoco rivers in South America and to African wetlands such as the Sudd and Inner Niger Delta.

THE DISCOVERY OF THESE VAST new peat stores provides important new impetus for the protection and more sustainable use of the world's largest tropical wetlands. To their ecological and social benefits, we can now add their value in keeping carbon out of the atmosphere. By Gumbricht's estimate, the world's stock of tropical peat could be 350 billion tonnes, making a global total of more than 800 billion tonnes – equivalent to twenty years of carbon emissions from burning fossil fuels. But if we want to keep that carbon buried, time is short. Because the drainage engineers are coming. Already, an estimated 500,000 square kilometres of peatlands, equivalent to the size of Spain, have been drained, mostly to create agricultural land.[221]

Peat soil in the Congolese Cuvette Centrale swamp, one of the most carbon-rich tropical regions in the world.

FOLLOWING PAGES:

A palm oil plantation at Igende, near the Cuvette Centrale peatland. Such plantations, which require the land to be drained, are a growing threat to tropical peat swamps from Peru to the Congo and Borneo.

But as the peat dries out, it oxidizes, releasing the carbon into the air as carbon dioxide. These drained swamps are probably already responsible for between 5 and 8 per cent of global greenhouse-gas emissions from human activity.

The biggest reason for draining tropical peat today is to cultivate oil palms. One of the world's most profitable crops, palm oil is used in a huge variety of consumer products from food to cosmetics. Over a typical twenty-five years of cultivation, a hectare of oil palm will emit more than 2,000 tonnes of carbon dioxide, according to Gumbricht's colleague Kristell Hergoualc'h.[222] Around half the world's palm oil

is currently grown in Indonesia. With roughly a fifth of that crop growing on drained peat, it likely emits around 500 million tonnes of carbon dioxide a year, more than half the country's total emissions.

The problem is exacerbated by fire. Farmers often burn forests to clear land for planting. Their fires often get out of control and ignite peat below ground. Fire is, in chemical terms, very rapid oxidation. During September and October 2015, burning forests and smouldering peat swamps in Indonesia added a billion tonnes of carbon dioxide to the atmosphere – more than the entire economy of the US.[223]

Climate scientists have pleaded with governments to halt the haemorrhaging of carbon from peat swamps. They say stopping global warming at 2° C will otherwise be impossible. They calculate that protecting the world's remaining peatlands could, by 2100, cut the likely concentration of carbon dioxide in the atmosphere by enough to shave a third of a degree Celsius off global warming.[224]

Draining peatlands also results in severe local impacts. The land surface typically subsides by between 3 and 6 centimetres per year in the tropics, massively increasing the risk of flooding productive land and permanent invasion by the sea in coastal areas.[225,226]

Clearly, urgent action is needed, and a start is being made. After his country's catastrophic fires in 2015, Indonesia's President Joko Widodo promised to protect his country's remaining peat swamps, and to staunch existing emissions by rewetting some of those lost. He created a Peatland Restoration Agency, charged with restoring 24,000 square kilometres[227] of burned and drained peatlands by 2020. The Agency began with a project to block 4,000 kilometres of drains dug back in the 1990s as part of the so-called Mega Rice Project in central Borneo. The project proved an agricultural disaster and produced not one grain of rice.[227] But the drained peat has been exuding carbon dioxide ever since. If all goes to plan, that will soon cease.

Staunching Indonesia's carbon dioxide flows will be a hard task. Despite Widodo's promises, many Indonesian companies claim they still have licences to drain more peat for both palm oil and forest plantations. But some of the big players, under pressure from consumers, do say they want to join the task. Asia Pulp & Paper, one of the country's big two, has begun retiring plantations on peatlands, and has been working with Greenpeace to figure out how to halt emissions from some 45,000 square kilometres of its peatlands across the country.[228] And there are efforts at 'South–South' collaboration between tropical peatland nations, including Indonesia and the Congos, to share expertise. Wetlands International and the Global Peatlands Initiative have been offering their expertise, too. ○

Western lowland gorillas in the peat swamp of the Cuvette Centrale in the Congo rainforest. As well as being a globally important region for carbon storage, the swamp is home to fourteen globally threatened species including bonobos, gorillas and chimpanzees.

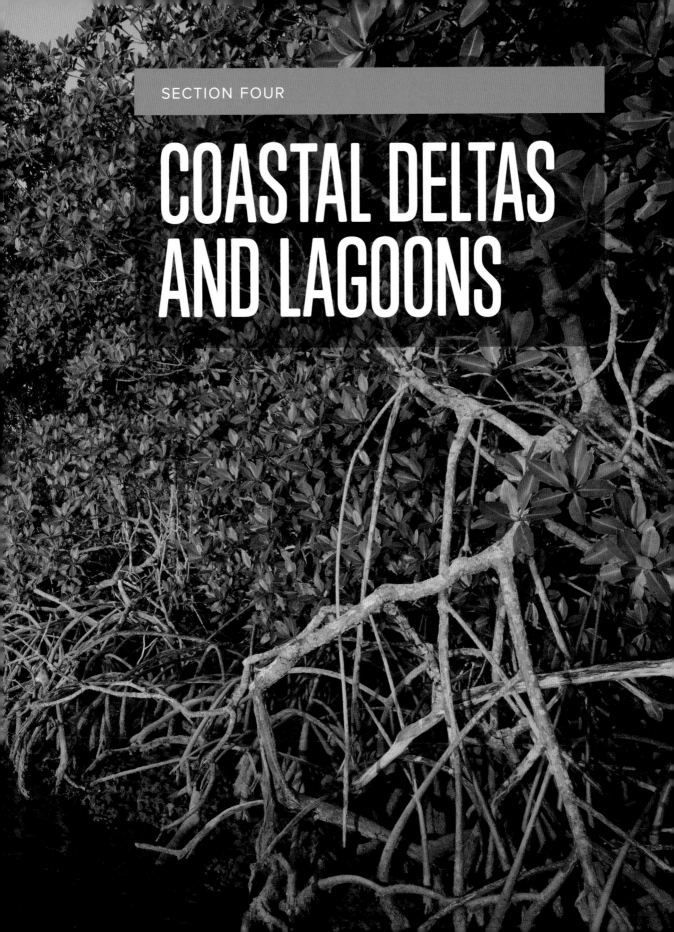

SECTION FOUR

COASTAL DELTAS AND LAGOONS

Most rivers reach the ocean through coastal wetlands, where fresh and salty water meet. These places have their own unique ecosystems. And whether mangrove swamps or salt marshes, tidal lagoons or river deltas, they absorb wave energy, take the brunt of winds, soak up tidal surges, capture sediment to build coastlines, and provide spaces where floodwaters can collect. Their loss has worsened tsunami disasters, and exposed millions to rising sea levels. A global spend of $10 billion a year on 'hard' coastal defences fails to hold back the tides.[229] No wonder that, from the Bay of Bengal to the Louisiana bayous and the North Sea, efforts to restore coastal wetlands are intensifying.

■
PREVIOUS PAGES:
Mangroves, such as these in the Everglades in Florida, are among the world's most productive coastal ecosystems, and hold more carbon than similar areas of rainforest.

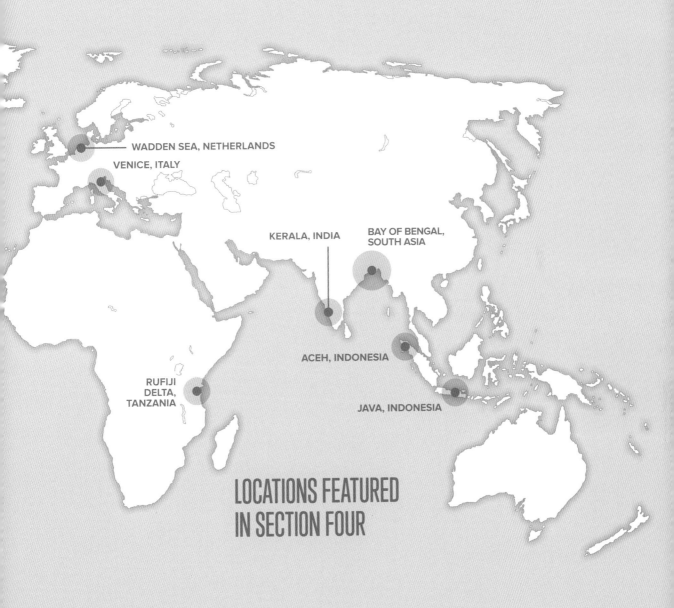

LOCATIONS FEATURED IN SECTION FOUR

- WADDEN SEA, NETHERLANDS
- VENICE, ITALY
- KERALA, INDIA
- BAY OF BENGAL, SOUTH ASIA
- ACEH, INDONESIA
- JAVA, INDONESIA
- RUFIJI DELTA, TANZANIA

RUFIJI DELTA, TANZANIA
RESTORING WETLAND RIGHTS FOR WISE USE

THE SAD-FACED YOUNG WOMAN, carrying a baby on her front, had been sitting quietly at the back of the meeting for some time, politely leaving the floor to the men of their village in the midst of one of Africa's largest mangrove swamps. Then she rose. Some of the other women at the back of the room looked surprised. But, with a government official in the room, she was determined to have her say. 'It is difficult to feed our families here,' she said, 'because we are short of food. We depend on rice and fish, but now there is no fishing allowed, and you want to stop us growing rice.' To buy food, the villagers were raising money by cutting nearby mangroves. But that was banned too, at least in theory. 'If we cannot fish, cannot grow rice, and cannot cut mangroves – we will have no income and no food,' she said. 'How will we live?'

Her short speech, in a meeting arranged for us in Nyamisati village on the delta of the River Rufiji in southern Tanzania, had been made with so little drama that her baby slept on. But it was firm. The village men mumbled their assent at her directness. Even the official, from Tanzania's Forest Service, which set the rules she complained about, could only agree. Government attempts to preserve the delta's ecosystems by banning locals from using its abundant natural resources made little sense.

The Rufiji is Africa's sixth largest river, behind only the Congo, Zambezi, Nile, Niger, and Volta. Its mangrove-covered delta on the shores of the Indian Ocean occupies an area rather bigger than the

Netherlands. Most modern readers will never have heard of this economically underdeveloped backwater. But that was not always the case. Search the internet for 'Rufiji Delta' and the first thing that comes up is the 'Battle of Rufiji Delta', a headline-grabbing skirmish in the First World War, when the most powerful German ship in the Indian Ocean, SMS *Königsberg*, was cornered there and sunk by British warships.[230]

Even before then, this delta had a history. What might today be mistaken for pristine swampland was intensely harvested in the nineteenth century, says Betsy Beymer-Farris of Furman University,

South Carolina.[231] The local people, known as the Warufiji, mixed fishing with mangrove harvesting and shifting cultivation of rice. For centuries, they traded coconuts, prawns, mangrove poles, and cashew nuts with Somalis, Persians, Omanis, Portuguese, Indians, and for a while the British. Fleets of Arab dhows were a common sight in the delta's waterways.

After the 1970s, the socialist Tanzanian Government made the traders unwelcome. One reason was to protect the delta's ecosystems from over-exploitation. But the Warufiji contest that. They said they had always managed and exploited nature on the delta, and should be allowed to continue. Outsiders who know the delta best tend to agree. 'Current land use ... is not nearly as extensive as it was during the eighteenth and nineteenth centuries,' says Beymer-Farris. Conservationists who want an unused landscape rather than a productive one, and who imagine that there was till recently a pristine delta, are guilty of 'misreading the Rufiji landscape'.[232] The delta is more used, and more resilient, than they think.

So, why did the government persist with bans on using the delta's abundant resources? Why the impasse between officials and villagers? How can a better, more sustainable and more productive future be developed amid the mangroves? The sad-faced woman, for one, would like to know.

THE RUFIJI DELTA TODAY is a backwater in every sense. Most buildings are made of mud, with thatch or tin roofs and no glass in their windows. Few villages have piped water. They depend on open wells that in some areas are increasingly salty. Most people have no electricity either, other than from the solar cells that a few of the better off have erected on their roofs. There are hardly any roads, and the dirt tracks are often impassable in the wet season. The inaccessibility exacerbates the lack of basic services. Some villages have a small clinic or a basic primary school. But for most people in the delta it takes hours by boat to reach a school or a doctor. No wonder literacy and life expectancy are low, even by Tanzanian standards. 'Very few people get higher levels of education, and when they do, they rarely come back,' one village chief said.

Efforts to change this situation have been fitful at best. The outskirts of several villages have rusting signs of long-forgotten and sometimes wildly misconceived development projects. In the 1990s, the government backed a plan to build the world's largest prawn farm. It would have privatized a third of the delta, and was abandoned after angry local protests.[233] More sensible was the Rufiji Environment

A fishing boat in Tanzania's Rufiji Delta.

PAGE 200:
The Rufiji Delta contains East Africa's largest area of mangroves. Long-established indigenous systems for managing the mangroves have been undermined by government attempts at conservation.

COASTAL DELTAS AND LAGOONS

Management Project, an initiative of the International Union for the Conservation of Nature, which drew up management plans to help four pilot villages manage their own natural resources. But the government was never keen, and the project abruptly closed in 2003.[234]

'So now two decades later, we have to start again,' says Julie Mulonga of Wetlands International, whose Mangrove Capital Africa initiative is another attempt at empowering Rufiji Delta communities. Her staff have begun working with the Tanzania Forest Service and local communities to revise a long-outdated mangrove management plan for the delta.[235] If all goes well, it could break the deadlock between the two sides. But will it? Can cooperation work?

To try and find out, we toured the delta with the Tanzania Forest Service's Regional Manager Mathew Ntilicha. He took an often conventional, confrontational view of his role as environmental policeman of the delta. He has 550 square kilometres of mangroves to protect, he said, but only five staff and a single fibreglass boat to aid him. 'We try to cooperate with communities, but sometimes we are fighting with them,' he said. A truck piled with confiscated illegally cut timber sat outside Ntilicha's office. His field staff told a similar story. 'The villagers do many illegal things. They believe the mangroves belong to them,' one said. 'They remove the mangrove trees, and because they don't have enough space to do their farming, they cultivate their rice in the mangrove areas. They run when they see us.'

Some of the confrontations have turned nasty. In 2011, armed police torched over 3,000 huts on rice farms in the north of the delta, as part of a forced evacuation.[236] It was clear that changing the relationship to one of cooperation would be difficult. Ntilicha said he would like to find a way of working with the delta communities. 'We want a collaborative plan to define areas for environmental protection, areas for cutting poles, areas for fishing, and areas for cultivation. The government cannot do it alone. We'd just have more confrontation. We have to engage by educating the villagers and giving them responsibilities. There can be cutting and selling of trees, but in a proper way.'

OUR JOURNEY THROUGH THE DELTA began in Nyamisati, a small market town of 2,600 people. It is the centre for many services for the delta people, and their main interface with the wider world. It has a secondary school for girls, electricity, a fresh water supply, and a mobile phone signal. The market sells rice and flour – though a big freezer store for fish, with a plaque celebrating its inauguration in 2016, was out of operation when we visited. There are bars where,

PREVIOUS PAGES:
The Rufiji River is Africa's sixth largest. The river's varying flows and constant supplies of silt from upstream maintain the mangrove swamps of its delta on the shore of the Indian Ocean. But a dam under construction upstream threatens to disrupt the delicate hydrology and undermine the delta's fertility.

some residents complain, drinks companies hold raucous promotions. 'Before, there wasn't alcohol here; now there is alcoholism,' said one.

From the town's small jetty, we headed through meandering waterways into the heart of the northern delta. Across wide areas, the mangroves are intact, lining the river banks and stretching across islands and peninsulas. Sediment brought down by the river is forming mud banks that are being colonized by new saplings. Somewhere in the forests lurk wild pigs, monkeys, baboons, warthogs, and mambas.[237] But in places we saw bare ground, where rice paddies had been abandoned, some following the government blitz on them a few years before. Straw huts where villagers had lived while tending their fields were charred, collapsing and falling into the water.

After two hours, we arrived at Mfisini, one of nineteen delta villages. Dozens of mature coconut trees were dotted among the mud huts, each tree owned by a different household. Chickens ranged free. In his wooden office hut, the village committee Chair, Yusuph Salelie, wanted to talk about the politics of mangroves. 'Everything in our lives depends on the mangroves,' he said. 'Our houses are built of mangroves; the fish we catch live in the mangrove roots; the mangroves clean our air; we even get salt from the mangrove areas.' The government has banned his people from harvesting mangroves in the interests of conservation. He agrees with conservation, but wants his village to regain the legal control they once had.

'The village used to issue a licence for people to cut the mangroves,' he said. 'We used to designate where and how many trees could be harvested.' Certain areas were protected by the Warufiji, who are predominantly Muslim. There were sacred groves and traditions and taboos about harvesting delta crops. 'It was all done in the village and it worked well.'

Sitting next to him, Ntilicha from the forest service looked sceptical. The government had banned harvesting the mangroves because 'the villagers had been taking things too far. There was no order.' But the ban was temporary. 'When we have a management plan, we will reinstate harvesting.' He pleaded for Salelie to cooperate in developing the plan. But he had a stick as well as a carrot. 'If you don't, it will force the government to keep the ban.' Salelie smiled. 'We all want to be part of the plan. We all plant mangroves and we all want to protect them.' Clearly there was common ground. Equally clearly, trust was the missing ingredient. Despite the friendly talk, it might all end in another round of arrests and hut burning. We headed for the southern delta to learn more.

The northern and southern halves of the Rufiji Delta are distinct. But they have been changing. The delta has a strong spring flood season, when freshwater comes down the River Rufiji. Until the 1970s, the largest volume flowed through the southern half of the delta. Then a series of unusually high floods brought sediment that largely blocked the river's southern route through the delta. The main channel shifted to the north.[238]

As a result, sea water penetrated further into the southern delta. Mangroves species best suited to salt prospered and rice cultivation, which needs freshwater, faltered. Villagers began shifting rice growing to the northern delta. That, villagers told us, is what caused the conflict with the government. Officials unfamiliar with the changes in the delta thought the clearance of mangroves to plant rice in the north was a reckless expansion of cultivation. In reality, the villagers were just adapting to changing environmental conditions.

We arrived at Ruma, a small village deep in the southern delta. Like all the villages we visited, it is well-organized. It has an environment committee, and villagers also elect resource groups to manage activities such as mangrove planting, fishing and bee-keeping. Those running these groups all said their authority was constrained by the laws imposed by the government. 'It would be a good thing if we had charge of the licensing. The mangroves are our resource, and we are the ones who would be affected if there are no trees,' said Mohamed Hamis, the Environmental Officer in Ruma. He admitted that the local mangroves suffer from people cutting them for sale to outside traders. 'The government cannot stop them,' he said. But if the village authorities had control, they could do the policing because they know who the perpetrators are. The village Chair, Kassim Rashid, agreed. Local control would 'improve livelihoods, including our fisheries, and help conserve mangroves for future generations'.

Our final stop was Jaja on the coast of the southern delta. For outsiders like us, it was the most remote village.[239] But, curiously, it felt more connected to the outside world with its air strip, hidden in the bush, its soccer pitch, TV dishes nestling beside solar panels on tin roofs and an old sign from a WWF fisheries project. A youth was riding a motor bike around the village and the forest sounds were interrupted by the dull thud of electronic music. We saw an old Arab dhow by the jetty, evidence of long links with Arab traders across the Indian Ocean that perhaps persist.

Unlike other more traditionally male-dominated delta villages, women play a leading role here. Jaja was the first, and is probably still the only village to have had as many women as men in its village

The mangrove kingfisher is found only in the mangrove swamps of East Africa, including here in the largest of them, Tanzania's Rufiji Delta.

government. The dominant character at our village meeting was a woman. Dia Kiyonga arrived late – deliberately so, it seemed, certainly unapologetically – and shook hands firmly with all the visitors. She is the village mangroves expert, she said, launching into a list of the six local species and an exhaustive description of their economic and ecological value to the community.

Most important is timber for house building. Their wood also kindles fires and stoves, builds beehives and fences, and provides poles for pounding grain. There are more specialized uses too, depending on the species, including medicines, dyes, and tools such as fishing floats.

With few roads, most travel within the Rufiji Delta is by boat. Even trucks go by water. Here residents wait for one of the regular ferries.

Not to mention alcohol from fermented sap. Their roots, meanwhile, are places where fish breed and feed, while preventing coastal erosion on the exposed Indian Ocean shore.

Far from losing their mangroves, Kiyonga said, Jaja had more than ever, 'because we are taking better care of them. We have an area where we plant them, where there is wet mud rather than sand.' What did she think of the government drawing up a new management plan for the mangroves? She waved away the very idea. 'We don't need a government plan,' she said. 'We have our own plan. I am in charge of it. I tell people where and when to harvest, or not.'

What should we conclude? It is clear that there is considerable potential for the resources of the delta to be used wisely and in a sustainable manner without destroying them. It has been done before. It is also clear that the Warufiji people are serious about managing their own environment, and generally know far more about it, and how it might best be harvested, than outside forest policemen, conservationists, or scientists. Likewise, there can be little doubt that, without their consent and active engagement, any attempt by outsiders to manage these resources will fail. Most obviously, nobody seriously thinks the bans on mangrove cutting or rice farming or fishing or anything else can be rigorously policed for long.

The challenge is to find ways to help the Warufiji improve their livelihoods without wrecking the ecosystem that underpins their lives. Wetlands International is working with delta communities on local methods of conservation and mangrove restoration and the development of a management plan that can benefit everyone.

THAT IS THE GOOD NEWS. Then came the bad news. Just as village leaders were signing up to take part in developing the management plan, reports spread across the delta of a new 'development' project. In late 2018, the government awarded contracts to an Egyptian firm for a 130-metre-high hydroelectric dam in Stiegler's Gorge on the River Rufiji, not far upstream from the delta. The dam would create the largest reservoir in East Africa, and hold back more than a year's river flow.[240]

It was a project that dam engineers have long dreamed of. It would at a stroke double Tanzania's electricity generating capacity. But ecologists have, for just as long, cried foul. They say it will wreck the Rufiji Delta's mangroves, fisheries, rice fields, wildlife – and ultimately the delta itself. Back in 1988, Raphael Mwalyosi, a young biologist at the University of Dar es Salaam, who later became the Chairman of the government's Rufiji Basin Development Authority, warned of the potential harm. 'Mangrove stands in the delta would probably be displaced by reeds. Delta fisheries would be very negatively affected, because of changes in the water regime as well as salinity levels,' he wrote. 'Floodplain fisheries would totally collapse.'[241]

He was not available for comment for this book, but that diagnosis of the dam project is still the view of many. Oxford University's Barnaby Dye, in a report for WWF in 2017, warned that the dam would capture most of the estimated 16.6 million tonnes of sediment that currently enter the delta each year. The result would be 'increased erosion, the isolation of lakes ... less soil fertility, the shrinking of the delta and reduced fish, shrimp and prawn fisheries'. It 'could negatively impact over 200,000 livelihoods'.[242]

The government's cursory environmental impact assessment of the project, submitted in May 2018 and seen for this book, agrees that a 'greater part' of the sediments brought downstream by the river would be trapped behind the dam, causing 'relatively strong degradation' of river banks downstream.[243] But it does not address the ecological or social consequences.

For its part, the government says that the dam will be good for people living downstream, because it will prevent floods. Ecologists disagree. They say that floods are just what the delta needs. They sustain it. Stéphanie Duvail of the Institute of Research for Development in Marseilles, and Olivier Hamerlynck of the National Museums of Kenya, two old Rufiji hands, say regular floods are 'essential for the sustenance of floodplain fertility [and] vital to the productivity of most of the natural resources on which local communities depend'.[244]

That is what we heard from delta villagers too. Floods could sometimes be a short-term inconvenience, they said, but the silt and nutrients they bring make bumper fish catches and maintain the fertility of the delta's soils. Far from being bad for people in the delta, floods are good. Sadly, in Dar es Salaam, such messages seem to have got lost. President John Magufuli told critics, 'Come rain, come sun, Stiegler's Gorge Hydroelectric Dam must be constructed.'[245] As good as his word, in late July 2019, he visited the gorge to lay the first stone.[246] ○

MISSISSIPPI DELTA, LOUISIANA
OL' MAN RIVER KEEPS ON ROLLING

ATTEMPTING TO TAME THE MIGHTY Mississippi has been one of America's most costly, longest-lasting and least-successful civil engineering endeavours. The first flood defences on the river were erected by eighteenth-century French settlers around the delta town of New Orleans. Later, slaves began raising levees for hundreds of kilometres upstream along the river. In 1879, the newly formed Mississippi River Commission brought in the Army Corps of Engineers to build yet higher levees. But the river kept hitting back. In 1927, it broke those new levees and submerged an area the size of Belgium. Hundreds died and the flood destroyed every bridge for 700 kilometres downstream of Cairo in Illinois. Billions more earth-moving dollars later, floods in 1973 breached 400 kilometres of levees. Much of St Louis was submerged and nearly 500 counties were covered by a presidential disaster declaration.[247]

Clearly the taming of ol' man river was not going well. One reason should have been clear. As the flood defences were raised, the most important flood defence of all – the river's floodplain – had been largely wiped out. Wetlands along the Mississippi that could once have stored sixty days of the river's water flows could, by the late twentieth century, manage only twelve days' worth. Preventing floods upstream had delivered flood disaster downstream. Mark Twain was surely right when he declared that 'ten thousand River Commissions with the mines of the world at their back, cannot tame the lawless stream, cannot say to it "Go here" or "Go there" and make it obey.'[248]

The story of the lost floodplains is one part of the troubled flood history of the Mississippi. It is essentially a repetition of Europe's experience with the 'rectification' of the Rhine. But there is another side. Decades of engineering the river also wrecked its delta. And that was especially bad news for the biggest city on the delta, New Orleans. Bad news that hit home when Hurricane Katrina came to call in 2005. For the enfeebled delta left the city fatally exposed to floods coming not from the river, but from the ocean.

The Mississippi is the world's sixth muddiest river. It carries plenty of material to make and maintain a large delta. For thousands of years the river's generous delta spread out into the Gulf of Mexico. But as humans began to canalize the river and rush its waters to the ocean, most of that silt stayed suspended in the fast-flowing river, bypassing the delta and ending up on the bed of the gulf. Suddenly starved of silt, the sea began eroding the delta. During the twentieth century, almost 5,000 square kilometres of mudflats and bayous were lost. The river entered the gulf instead along a narrow isthmus of land, held in place by levees, while all around the former delta was flooded. They call the delta the 'crow's foot', these days, because after decades of erosion that is what it now resembles.

As the delta has disappeared, dozens of bayou communities have lost their farmland to the sea. Many have become isolated clumps of houses on scraps of protected land surrounded by water.[249] The fishing community of Leeville has lost two-thirds of its wetlands, and its cemetery can now only be reached by boat.[250] In 2016, the tribal settlement of Isle de Jean Charles received a federal grant to relocate.[251]

The creeping disaster of the disappearing Mississippi Delta reached public attention outside Louisiana when Hurricane Katrina swept ashore from the Gulf of Mexico in 2005. Across the delta, hundreds of square kilometres of land disappeared under water and never re-emerged. The hurricane's winds pushed a tidal surge up a shipping

New Orleans blamed its 2005 flood disaster on Hurricane Katrina. But the real cause was the destruction of the wetlands that would have saved the city.

canal through the delta right to the gates of New Orleans, the delta's great city.²⁵² Even with three centuries of flood defences, it stood no chance. The waters breached the defences of New Orleans in fifty places, flooding 80 per cent of the city and killing 1,800 people, almost half by drowning.

Following the catastrophe, Louisiana's coastal engineers belatedly saw the light. They no longer boast of their ability to hold back the waters, whether from the river or the ocean. Instead, they forecast that a further 5,000 square kilometres of the delta was likely to be lost over

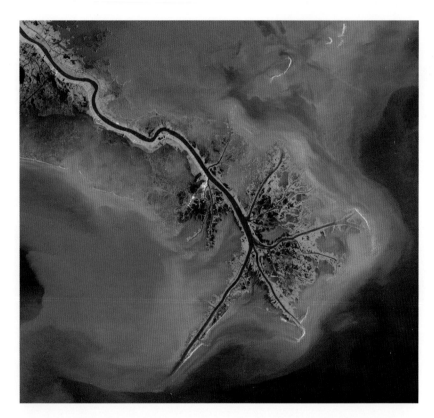

LEFT: The distinctive 'crow's foot' shape of the outer delta of the Mississippi. One of the world's longest rivers, it was formerly one of the muddiest, too. Flood defences upstream have deprived the delta of silt and triggered massive erosion by the sea, leaving the delta's crow's foot stranded. One outcome was the great flood in New Orleans in 2005.

OPPOSITE: Dredging of shipping lanes in the Mississippi Delta has increased its vulnerability to invasion by the waters of the Gulf of Mexico.

the next fifty years, and came up with a post-Katrina master plan to go back to nature to rebuild the delta. Their thirty-year plan, published after long consultation in 2017, proposes protecting existing marshes, while dredging material from across the delta to create 1,800 square kilometres of new wetland – equivalent to around a third of what has been lost since 1930.²⁵³

The plan should create habitat for everything from alligators to oysters, while restoring offshore barrier islands and creating new routes for the river to cross its delta. The shipping lane that funnelled Katrina's flood surge into New Orleans will be closed. Some communities and much land will be sacrificed to save others and to help the natural wetland systems rebuild. Next time a hurricane comes calling, they hope that nature will absorb the hit rather than the people of Louisiana.

WHAT HAPPENED ON THE MISSISSIPPI DELTA in the twentieth century has been happening on other river deltas around the world. Dams and dykes have starved them of silt and water from upstream. In Pakistan, the town of Keti Bunder stands on the edge of the

disintegrating delta of the River Indus. It was once the centre of a rich farming area. The government had planned a major port.[254] Now it is just a few mud huts and wooden shacks surrounded by earth banks to keep the sea away. Its residents need water delivered by tanker, since their own supplies are too salty to drink. The surrounding district once had forty-two settlements, but twenty-eight have been lost to the sea.[255]

The Indus used to be a mighty river, flowing out of the Tibetan Plateau, through the Sindh Desert of Pakistan to the Arabian Sea. It had twice the volume of the Nile. The river constantly resupplied its 40,000-square-kilometre delta, the world's sixth largest, with eroded material. But no more. Thanks to British colonial engineering, most years the waters of the Indus are largely diverted to irrigate an area of Sindh that is bigger than England. The delta receives only a tenth as much water as it once did, and a quarter as much silt.

Unsurprisingly, the delta is in full retreat. Half its mangroves have washed away since the 1950s.[256] A replanting programme largely failed as scouring waves consumed the saplings. Farmers have become fishers and freshwater fishers have become sea fishers. Many delta residents have become environmental refugees, heading for Karachi, the fast-growing megacity on the edge of the delta.[257] But just occasionally, heavy monsoon floods rush out of the mountains. It happened in 2010, when a fifth of Pakistan was under water. But while the rest of the country suffered, for many of the delta's inhabitants it was a time of rejoicing. As the creeks filled with freshwater, and silt washed across the fields, farmers planted once again. As mangroves revived, fish returned.[258] One delta fisherman told journalists, 'We've been waiting for this water for the past fourteen years.'

Dams and river diversions have been a disaster for deltas. A quarter of all the sediment travelling down the world's rivers is now trapped behind dams or ends up in irrigation canals.[259] Since the completion of the High Aswan Dam on the Nile in Egypt in 1970, most of the river's 100 million tonnes of silt has accumulated on the bottom of the dam's giant reservoir. As a result, the Nile Delta – which sustained agriculture for more than 5,000 years and is home to two-thirds of Egypt's 100 million people – has retreated by more than 2 kilometres, drowning several villages.[260]

The delta of the River Colorado in North America has fared even worse. A hundred years ago, it was a rich haven for wildlife in the Sonoran Desert, sustained by sediment gouged from the land during the creation of the great canyons along the river's route from the Rocky Mountains. Ecologist Aldo Leopold wrote of 'a verdant wall of mesquite and willow ... a hundred green lagoons ... the river was everywhere and nowhere'.[261] It was a place where jaguars roamed and beavers swam. The Cucapá people lived on the delta's islands and cruised its waterways in dugout canoes, hunting wild boar, harvesting wild grain, and cultivating beans and squash. Paddle steamers travelled through the delta to the ocean.

But during the twentieth century, engineers built ten major dams and more than eighty water diversions, to generate hydroelectricity, irrigate thirsty crops in California and water the golf courses of desert cities such as Las Vegas, Phoenix, and San Diego. Deprived of its

White pelicans on the Mississippi. The birds forage in the river's shallow marshes and lake edges.

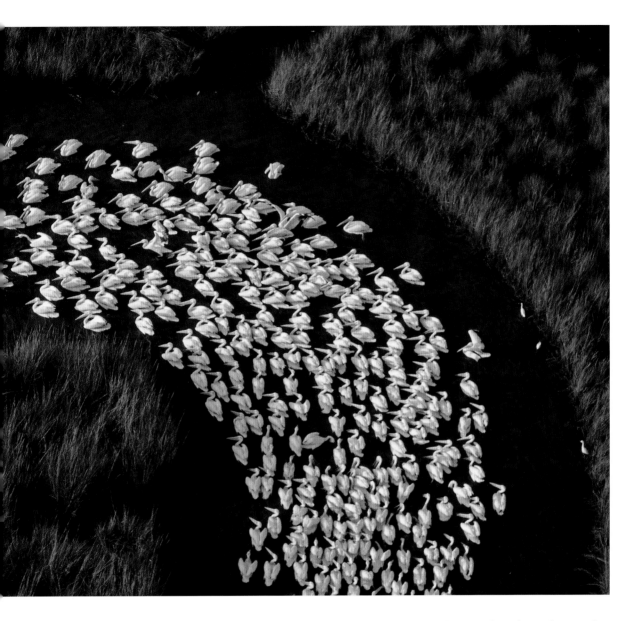

lifeblood, the once-magnificent delta has been reduced to a few pools and brackish mudflats, sustained by agricultural drainage.[262]

Things could change, however. The dams are much less important to American electricity supplies and food production than they once were. They could be managed to resume supplies of water and silt to the delta, or could one day be torn down altogether.[263,264] If that happened, the delta's ecosystems would come back. We briefly saw what was possible in 2014, when the river's managers ran an eight-week experiment, releasing 130 million cubic metres of water from the Morelos Dam to create an artificial spring flood in the delta.[265] Willows and cottonwood trees germinated, and a green flush spread across the barren delta. Since then, an agreement between the US and Mexican governments to buy back the water rights of some farmers has allowed small continuing flows, aimed at rewetting a few delta restoration sites. It is a small start, but it is generating an appetite for more.

SOMETIMES HUMAN ACTIVITY can increase the silt loads of rivers, and expand their deltas. Many of New England's coastal marshes expanded after European settlers began logging the land upstream.[266] A study of four of southern Europe's largest river deltas – the Ebro, Po, Rhône and Danube – found that they all expanded from the seventeenth to the nineteenth centuries, when population growth led to deforestation and erosion of their catchments upstream.[267] During that time, the fertile knob of Spain's Ebro Delta poked ever further out into the Mediterranean. There was so much silt that rice farmers could buy extra with their water supplies, as a handy fertilizer, says Carles Ibanez of Catalonia's Institute of Agri-food Research and Technology.

That growth abruptly ended in the twentieth century, when dams on the rivers began to trap sediment. For more than half a century now, all four deltas have been in retreat. Since the completion of the Mequinenza Hydroelectric Dam on the Ebro, the annual silt input to the river's delta has fallen by 99 per cent.[268] The delta has been retreating by up to 50 metres a year. Now, the rice fields are filling with salt from the sea. 'The locals keep calling for hard protection, but what the delta really needs is more sediment,' says Ibanez.[269]

That is not an impossible dream. Ibanez says it could be achieved by opening the dam gates for a few weeks each year during periods of high rainfall. A fast flow could flush out sediment that has accumulated on the reservoir floor. Such a project would be good for the dam's owners, too, he says. Because the flushing would increase the reservoir's water-storage capacity, and hence its power-generating potential. Other dams and river deltas could benefit from a similar strategy. As on the Colorado, the lesson is that dams need to be used more creatively. The control the structures have over rivers and their wetland ecosystems is huge, and till now it has been largely destructive for downstream nature. But that need not remain the case. They can be used for ecological good as well as ill. o

KERALA, INDIA
CRAZY BACKWATERS IN 'GOD'S OWN COUNTRY'

IN INDIA, THEY CALL THE SOUTH-WESTERN state of Kerala 'God's own country'. That was not how it felt in August 2018, when monsoon floods devastated its densely populated low-lying coastal plain. Around 500 people drowned in an area best known to outsiders for its placid backwaters, where tourists' houseboats cruise for days across lagoons and down picturesque canals. As the floodwaters abated, questions were being asked about whether the disaster was made worse by water engineering projects in the backwaters. Schemes designed to feed the state's population and pamper tourists may have both generated flood surges and damaged the ability of the backwater wetlands to cope with them. Was 'God's own country' damned as well as dammed?

The floodwaters came out of the Western Ghats. The chain of mountains down the west side of India is one of India's wettest places, drenched from June to September in monsoon rains. In early August

2018, those rains were exceptionally intense and unremitting. Rivers out of the Ghats dumped their water into the backwaters on a coastal plain that is often below sea level. The 100-kilometre-long Lake Vembanad, which is at the heart of the backwaters, rose up and engulfed surrounding wetlands, rice fields, farming villages, and cities. A quarter of a million people took refuge in relief camps; 10,000 kilometres of roads and 300 square kilometres of farmland were damaged; and Cochin airport was awash in muddy water.[270]

Four months on, when we visited, the clean-up had been largely completed in many places. Often, the only visible reminder was the tide mark on buildings. But in some of the poor rural communities hardest hit, recovery had barely begun. At Kannady, a small village in the Kuttanad Wetland south of Lake Vembanad, the Red Cross only arrived in early December to offer tents to families whose houses had been destroyed. Its people still slept in broken tin shacks, before spending their days laboriously digging drains across their rice fields to flush out polluted water. Only then could they plant a new crop to feed their families.

Some Kannady villagers blamed the floods on the backwaters. After all, as village Councillor Ambila Gose pointed out, the water had come up from the lake and across the Kuttanad Wetland, before swamping their community. Maybe more drains would empty the wetland and keep the water away, they suggested. But the truth is the opposite, said B. Sreekumar of the Kottayam Nature Society, which is based on the wetland. Two-thirds of the wetland had been drained by farmers in the past century. Those wet areas that survived had minimized the flood damage by absorbing floodwaters. What the villagers need is more wetland, not less. 'The flood is man-made. To reduce the flood risks, we should remove all the encroachments,' he said.

The backwaters that dominate the narrow coastal plain of Kerala are a delight for tourists. Millions of Indians and foreigners come each year. Some now head for locations featured in Arundhati Roy's Booker prize-winning novel *The God of Small Things*.[271] Our boatman on Lake Vembanad claimed to have been brought up in the same village as her. It might even be true.

But the backwaters are not what they were. One reason is the tourists. Many of the mangroves that once lined Lake Vembanad have been removed to make way for tourist lodges, and to improve their views of the water. Another reason is feeding India. Half a century ago, newly independent India rushed to drain the rich wetland soils of the Kerala coastal plain for growing rice. Dutch engineers brought their techniques of land reclamation to help. They converted parts of Lake

Wading through flooded streets in Kuttanad during the 2018 disaster on the backwaters of Kerala. Around 500 people drowned, in part because many of the wetlands have been drained.

FOLLOWING PAGES:

Locals transporting harvested reeds on the Kerala backwaters. As well as being a prime tourist destination, the wetland provides a natural bounty.

Vembanad into more than a thousand polders, each with stone walls surrounding drained land that is today a metre or more below the level of the lake outside. They also built a barrage across the lake to keep ocean salt water away from rice fields. Thanks to these measures, more than 40 per cent of the state's rice is now grown on wetland areas.[272]

Urban developments grabbed the backwaters too. In Kochi, the state's biggest city and main port, a few scraps of mangrove stands on river banks remind visitors of its wetland past. On the edge of Thrissur, another fast-growing metropolis, real estate developers have

just completed Sobha City on land annexed from the Kole Wetland on the north shore of Lake Vembanad. The large estate of high-rise apartments is surrounded by reed beds swaying in the breeze.

Urban construction has also triggered an explosion of sand mining in the rivers that feed the backwaters. In recent times, 12 million tonnes of sand a year has been taken from the catchment of the lake, a rate of removal forty times faster than the rivers replace it. River beds have been lowered by around 2 metres.

What remains of the Kole Wetland is still rich in wildlife. A recent bird survey found 250 species, said local environmentalist Manoj

Karingamadathil. Many were logged on a flooded patch of abandoned paddy near the city of Palakkad, where we saw a greater spotted eagle within seconds of our arrival. A pair of hoopoes crossed the path. There were marsh harriers and pelicans feeding. As evening drew on, local fishers joined them. But a local farmer told us, 'I don't like the birds. They eat my seeds.'

Despite such green enclaves, much has gone. By the end of the twentieth century, Lake Vembanad had halved to less than 180 square kilometres. With siltation reducing its depth, the lake's volume had diminished by three-quarters, according to Ritesh Kumar, Wetlands International's South Asia head. Meanwhile, fish stocks have suffered from the lake's barrage, which prevents migration into and out of the lake. And a combination of pollution and stilled waters has caused a plague of water hyacinths. The tourists' houseboats spend much of their time ploughing through mats of these weeds rather than cruising open water.

Whatever the ecological losses, engineers believed they had the complex hydrology of the backwaters under control. That their dams, dykes, and barrages could prevent floods as well as maintain the productivity of the fields and generate hydroelectricity. But that complacency was shattered by the 2018 floods, says Kumar. The system may be able to handle regular monsoons. But in exceptional years, the engineering has primed the coastal plain for maximum flooding.

One target of public concern during the floods was the dams that barricade many of the rivers flowing from the Western Ghats into the backwaters. Hydrologists pointed out that virtually all their reservoirs were full at the start of the floods.[273] They had no room to absorb the heavy rains coming downstream. 'Steps could have been taken to avoid the calamities downstream,' hydrologist E. J. James, a former member of the Kerala Dam Safety Authority, told the *Deccan Chronicle* at the height of the emergency. 'There were predictions about incessant rain and [the] water level was bound to increase.'[274]

As soon as the forecast came through, dam managers should have begun gradually emptying the reservoirs to create space. Instead, they wanted to keep their reservoirs as full as possible, so they would have water to generate electricity in the long dry season to come. Then, with their structures facing destruction from the force of the water behind them, they were forced into rushed releases. At the 138-metre-high Cheruthoni Dam on a tributary of Kerala's biggest river, the Periyar, operators ended up opening all five gates, for the first time in the forty years of its operation. 'The dams were never meant to be opened like that; it caused massive flooding downstream,' says Kumar.

Waterways in the picturesque backwaters of Kerala have become clogged with invasive water hyacinth from South America. This has followed the draining of marshes that is also raising the risk of floods during heavy monsoon rains, as happened in 2018.

The swollen river gushed across the Kole Wetland, inundating Cochin International Airport for several days. Thousands of people had to be evacuated.

Some dam engineers question how much difference the rushed last-minute opening of the dam gates made. K. P. Sudheer of the Indian Institute of Technology Madras, in Chennai, calculated that earlier releases from dams on the Periyar would only have reduced peak discharges by a fifth.[275] People in Kannady village questioned that, remembering how their floods dramatically worsened in the hours after the dam had been opened.

The debate about the operation of the dams has diverted attention from a more fundamental issue, says Kumar. Whenever the water was released from the overloaded dams, it had to go somewhere. But there was no longer anywhere it could be safely accommodated. The lakes and wetlands that would once have absorbed the flood have been dyked and drained to make rice fields, industrial estates, tourist resorts and cities. 'The areas where wetlands reclamation had been most significant were also the ones to bear the severest of the impacts,' he notes.[276]

We saw this ourselves in the village of Pullu, south-west of Thrissur on the River Manakkody. The village is an island surrounded by rice fields created from the Kole Wetland. Villagers told how the water came from the river, through their fields and into the village, reaching shoulder high in places. Some residents were rescued by helicopter. Many houses were wrecked. The village lost almost all its livestock, too. 'We released the cattle when the flood came, and not many returned,' said villager Siva Dason.

Four months on, they were recovering. They had pumped polluted floodwaters from their wells, and were rebuilding their homes. But first they had to eat. Badruddin, whose house had been destroyed, was out fishing in the river. And rice fields where the villagers had lost one rice crop to the flood, were already planted with the next.

It was moving to see their efforts to put their lives back together. But the truth is that the rice fields should not have been there. Officially, the land around the village is part of the Kole Wetland, which is recognized by the Ramsar Convention as an internationally important wetland. We asked villagers if they thought that draining the wetland might have left them vulnerable when the water came. The village leader K. Parameswaran thought for a while. Perhaps other communities were guilty, he said. It sounded like obfuscation, but it was a fair point. No village can save itself alone. It was a collective failure – a tragedy of the hydrological commons. The problem lies in attempting to turn large areas of natural wetland and river floodplain into dry land, says Kumar. 'This whole coastal area is a floodplain. It is where the water goes; it is natural. We should not try to prevent it.' o

VENICE, ITALY
WHEN A COASTAL LAGOON BECOMES AN OCEAN BAY

St mark's square was a lake. Dozens of tables set out for tourists had been washed away. The Basilica was flooded for the first time in a century. Three-quarters of Venice was reportedly under water in late 2018.[277] It was the worst invasion in half a century by the lagoon that makes the city unique. But it wasn't a one-off. The city is flooding ever more frequently. The reason is that land reclamation, dredging for ships, and the loss of silt due to river diversions, are all undermining the hydrological processes that the lagoon requires to protect the Italian city.

Venice is built on more than a hundred tiny islands, set in a large lagoon of tidal mud banks and salt marshes. The lagoon is separated from the Adriatic Sea by sand bars, with inlets through which the sea gently flushes the lagoon. Italians first moved there almost 1,600 years ago. After the collapse of the Roman Empire, the islands provided a refuge against Goth horsemen invading across the plains of northern

Italy. The lagoon was too shallow for invading ships and too deep for marching armies. The new colony persisted, and eventually grew into a great port, trading with the Orient. Its wealthy merchants built an architectural marvel and filled their waterfront houses with art treasures from around the world.[278]

More recently, those treasures have made the city a magnet for tens of millions of tourists each year. Venice is probably the most visited lagoon in the world. But the majority of visitors have eyes only for the city and its canals. If they notice the lagoon lapping around it,

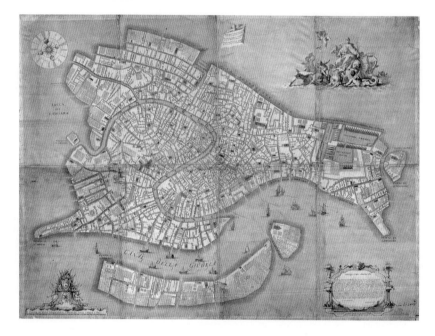

LEFT: Venice, as mapped in around 1730 by Italian engraver Giuseppe Baroni.

RIGHT: Rising tides, dredging, land reclamation, and the diversion of rivers have turned the sleepy Venice Lagoon into an ocean bay, scoured by the waves and in constant danger of being engulfed.

they mostly see it as a threat. Few realize that without the 550 square kilometres of shallow waters round it, there would be no Venice.

Almost from the earliest times, Venetians feared the complex hydrology of the lagoon, and began meddling with it. In the Middle Ages, they worried that the three rivers that entered the lagoon – the Brenta, Piave and Sile – were filling it with silt. So, around 600 years ago, their engineers began to divert the rivers directly into the Adriatic. But they overdid it. The diversions starved the lagoon of sediment. The mud banks, salt marshes and sand bars that protected the lagoon's shallows from the forces of the Adriatic Sea outside, began to disappear. The sea began to invade.[279]

At first, this invasion was slow. But it accelerated in the nineteenth century after engineers replaced the twelve narrow inlets to the lagoon with three wider, deeper channels that could accommodate

bigger ships. Incoming tides became higher, and sea currents scoured the mud and sand, making the lagoon ever deeper. It became less like a lagoon and more like a sea bay.

During the twentieth century, things went from bad to worse. A third of the lagoon's mudflats were drained and dykes for farming, the city's airport, and an industrial centre. This caused water levels everywhere else in the lagoon to rise, especially at high tide. Meanwhile, the islands sank, as Venetians pumped water from beneath them to supply industry. Today, they are about 12 centimetres lower than 150 years ago.[280]

What remains of the lagoon is still a prized wetland ecosystem. In winter, it hosts more than 100,000 water birds. Europe's largest concentrations of little and black terns converge here. Its fishery remains productive. Sea bass, bream, sole, mullet, crabs, squid, and eels enter the lagoon each spring and grow fat in the rich, silty waters. When they head back to the Adriatic each autumn, they must pass through traditional fishers' traps that harvest the big fish while allowing the youngsters to depart.

Nonetheless, the lagoon has been transformed. It contains twice as much water as it once did – and much less mud. Once, the lagoon was rarely more than 1 metre deep. Its filigree of creeks meandering through mud banks and salt marshes absorbed the force of waves coming through the inlets, and distributed the high tides. Today, the lagoon is in places 20 metres deep. Waves rush in, surging down the deepened shipping channels, scouring and eroding as they go. High tides regularly wash over the city's lower areas.[278]

Faced with this crisis, the city authorities still look to engineering. They have insisted that the only sure way to save Venice is to build a physical shield against the waves. They have spent $6 billion to install seventy-eight hollow steel barriers on the sea bed at the lagoon's three inlets. The barriers, each 20 metres square, are filled with water to hold them down. The plan is that whenever water levels in the lagoon are forecast to rise to a metre or more above normal levels, the water will be pumped out of the barriers, which will rise to block the entrances to the lagoon. After countless delays and a corruption scandal that forced out the city's mayor in 2014, the barriers may finally be operational by the time this book is published.

At first, the barriers will be expected to operate about seven times a year. But with half a metre of sea-level rise likely within a century, the lagoon could one day be shut off from the sea for more than half the year. That, say ecologists, would deplete it of oxygen, and make it uninhabitable for fish and birds. Not to mention creating a stink for the tourists.[281]

Rescue near Rialto Bridge: Tourists and the people who service them, regularly have to escape flooding across the low-lying parts of Venice.

FOLLOWING PAGES:
St Mark's Square, Venice's most famous landmark, is flooded several times a year as the lagoon invades.

The scheme's many critics say there should be a better way of saving Venice. They argue that centuries of engineering in the lagoon have undermined its natural defences against the sea. It has been transformed from a shallow, brackish coastal wetland that protected the city, into an open bay that is being swamped by the sea. What is needed, they say, is to rebuild those natural defences. 'Restoring the salt marshes is key to the resilience of the entire lagoon system,' says Jane Da Mosto, an environmentalist based in the city for many years. By giving water more places to go, the marshes would reduce

high-tide water levels throughout the lagoon, even as sea levels outside rise.

In places, this restoration is happening. Land that was reclaimed for industry in the twentieth century but never used is being reflooded. But much more is needed. Dredging of the shipping canals must be halted so that they can partially refill, Da Mosto says. And the gaping-wide inlets to the lagoon must be narrowed. Ultimately, too, restoring the lagoon will require the return of the rivers that flowed through it before being diverted. They would increase the natural supply of sediment necessary to fend off rising tides.

There would be opposition to all this, of course. The shipping lanes have boosted the city's economy by bringing in oil tankers, container ships, and cruise liners. But heavy industry is in decline, and tourists might reasonably be persuaded to forgo their grand entrance in return for saving what they came to see. Maybe rising sea levels outside the lagoon will mean that Venice will need its steel barriers too. But a restored lagoon would need them much less often.

Above all, a new mind-set is required. 'Engineers still see the lagoon as the enemy. But it is Venice's best defence against the sea,' says one

city architect, who has spent his entire career repairing buildings damaged by the repeated floods. The answer to Venice's crisis lies not with concrete and steel, but in a return to the soft engineering of nature, to the subtle interplay of silt and water across the ancient lagoon.[278] 'The city can only be saved by saving the lagoon.'

THE FLOOD MANAGEMENT of many urban areas around the world requires a similar change in outlook. Most modern cities could hardly be better designed to create floods. Many are built on low-lying coastal marshes which have been drained and dyked. Large expanses of hard surfaces mean rainfall flashes across the surface into rivers, rather than percolating into soils. Meanwhile, those rivers are tightly constrained, concentrating floods from upstream and from within the cities. Often, pumping of underground water to keep city taps running has also caused land subsidence. And coastal wetlands such as salt marshes and mangroves, which should be the land's first line of defence against the tides, have been reclaimed for real estate.

When Manhattan was first settled by Dutch fur traders in the seventeenth century, it was a narrow spit of dry land surrounded by salt marshes and mudflats.[282] As the island grew into the heart of the world's premier city, New York, those wetlands were slowly drained and claimed. Parts of Queens, Staten Island, The Bronx, and Brooklyn have always been vulnerable to floods. But it was a shock when superstorm Sandy inundated low-lying areas of Manhattan in 2012, humbling the human-made flood defences of some of the world's priciest real estate. Sandy caused damage costed at $15 billion in New York City alone. More than half a million homes were damaged. But it was mostly just nature reclaiming what had been its own.

In the years since Sandy, there has been a flood of ideas to make the Big Apple more resilient to floods. Most involve making it more absorbent. New York architect Stephen Cassell suggests recreating some of the marshes that once flanked Manhattan.[283] In places, this could be done by removing flood defences around existing open ground. Targets include Battery Park on the southern tip of the island, from where ferries depart for Staten Island, and Riverside Park on the shore of the Hudson River on the Upper West Side. Elsewhere, Cassell proposes extending the land onto the foreshore. In most cities, such land reclamation from the sea is zoned for high-value property; but here it would create space for marshes to help protect existing real estate.

Meanwhile, Cassell says, streets and other hard surfaces should be reconstructed to mimic some of the features of natural wetlands. Porous streets, sidewalks and parking lots would allow rainwater to seep underground and flow towards the coastal marshes. 'Man pushed nature out of most of Manhattan,' says Cassell. 'Maybe it is time to bring some of nature back, for our own good.'[284] Initially his ideas gained ground with city authorities. A plan was developed for an East River Park that could partially flood during storms, but they have since reverted to a more conventional raised sea wall.[285]

Even so, the desire to protect urban areas from floods by bringing back wetlands is growing. Just as rural landscapes are giving land back to fluvial systems, so urban areas can benefit from ceding ground to wetlands and giving rivers room. In 2018, World Wetlands Day

A clam fisherman working in the Venice Lagoon. Despite the hydrological damage it has suffered, the lagoon remains a fertile breeding ground for sea food.

COASTAL DELTAS AND LAGOONS

was devoted to urban wetlands. Its wish list included permeable pavements, reviving remnants of former wetlands never drained for development, creating new artificial wetlands, and even cherishing accidental wetlands on abandoned land.[286] The day's organizers at the Ramsar Convention said such initiatives would prevent floods, optimize local water resources, act as air purifiers, provide green spaces for recreation and wildlife, and become a cooling antidote to the 'heat islands' created by miles of concrete and asphalt.

Many cities have been recreating wetlands for public amenity and wildlife conservation. Since the 1970s, municipalities around San Francisco Bay have plugged drains that were installed on the bay's salt marshes and mud banks in the mid-twentieth century for building projects that never happened.[287] The aim is to reinstate habitat for iconic bay birds such as the California clapper rail. Southwest London's Barn Elms Wetland Centre has reclaimed for nature 30 hectares of abandoned water-supply reservoirs on a meander in the Thames. And, not to be outdone, north-east London claims to have Europe's largest urban wetland, on 210 hectares of former wasteland in the Lee Valley.

Elsewhere wetlands are being restored for more utilitarian purposes. Thirty Chinese cities agreed to create wetlands that can absorb 70 per cent of storm-water run-off by 2030.[288] Wetlands International has been advising Panama City on how to protect and restore wetlands to mitigate flooding caused by a construction boom in the Central American capital.[289]

In 2015, Panama's President Juan Carlos Varela reversed his predecessor's plans for building on an 850-square-kilometre mangrove-covered stretch of coastline that was a wintering stopover for migrating birds.[290] The aim was partly conservation: to protect the country's dwindling stock of mangroves, half of which disappeared between 1969 and 2007, and to retain habitat for anteaters, tapirs, loggerhead turtles, and a million migrating birds. But it was also intended to protect the city from floods. Low-lying areas that should benefit include the city's international airport and fast-growing suburbs such as Juan Diaz which lie on a river floodplain right behind now fragmented areas of coastal mangroves.

Mangroves, the world is learning, are among the most effective natural means of absorbing floods. o

ACEH, INDONESIA
MANGROVES TO FIGHT THE NEXT TSUNAMI

INDONESIAN FISHER HAJAMUDDIN was at sea the day the 2004 tsunami hit. It was the safest place to be. He only felt a mild swell as the wave generated by an earthquake beneath the Indian Ocean headed towards the shore of Aceh, the northernmost province on the Indonesian island of Sumatra. But when he returned to his home port, the coastal village of Gle Jong, he found it obliterated by the giant wave and under 3 metres of water. A village that had been home to 800 people was an open bay. Just seven residents survived the giant wave that hurtled up the beach that morning. 'My family was all gone,' he recounted ten years later. The lucky few were collecting firewood and had time to rush up the steps of the village's only high point which, with heavy irony, was its cemetery. There, they sheltered among the corpses as their families drowned.

Gle Jong was one coastal community among hundreds washed away on Boxing Day 2004, when the tsunami battered coasts for thousands of kilometres. An estimated 167,000 people lost their lives in Aceh alone.[291] Another half a million lost their homes. A decade on, the village was on the mend. 'People here are still traumatized. The faithful lost their faith,' Hajamuddin admitted. Survivors were still frightened by a glimpse of the ocean or the sound of wind in the trees. But an influx of newcomers such as Hajamuddin's new wife, and a post-disaster baby boom, had brought the village's numbers back to 130. They all live in new homes built with international aid and set further back from the coast.

The recovery was a remarkable story of human resilience. But it was more than that. For the community was doing its best to ensure that, should another tsunami hurtle towards their village, the coastline would be better protected. A few yards inland from the new post-tsunami shore, on land left waterlogged by the killer wave, the survivors have planted 70,000 mangrove trees. 'When the floods come again, the mangroves can save us,' said Hajamuddin.

Wishful thinking? Maybe not. Research into the tsunami tragedy has shown that a key factor in who lived and who died in coastal communities was whether the natural mangroves that once lined the shore were still there. Often, communities had removed them to make way for prawn ponds, rice paddies, or to make charcoal. But those places that still had mangroves are widely reported to have suffered much less, because the dense foliage and roots dissipated much of the tsunami's energy.[292]

Such anecdotal conclusions are hard to test in the field, obviously. But they are backed up by some research. One modelling study mimicking a smaller tsunami in Papua New Guinea in 1998, concluded that a 100-metre belt of mangroves reduced the destructive force of the wave by as much as 90 per cent.[293] Another estimated such a belt could reduce wave height by between 13 and 66 per cent.[294] Of course, thousands would have died regardless in Aceh. Many mangroves were themselves destroyed by the Boxing Day wave. But on the worst-hit western coast of Aceh, field research found that forests in front of villages reduced casualties by an average of 8 per cent.[295] That figure sounds modest, but on the fateful day it could have represented 13,000 lives saved.

Drawing on such findings, the calls to restore the coastal mangroves grew. Initially, to put back the 300 square kilometres of mangroves destroyed or badly damaged by the tsunami. And then to plant more on some of the 600 square kilometres of land permanently swamped by the floodwaters and on land which subsided due to the quake.

The Boxing Day tsunami in the Indian Ocean in 2004 blasted the shore of the Indonesian province of Aceh. The damage and loss of life was made worse by the removal of coastal mangroves to make room for fish ponds.

Many early projects failed. Most of the mangroves died through poor choice of species and location, or lack of aftercare.[296] Learning the lessons, Wetlands International and Oxfam Novib launched a new initiative. Rather than just paying for villagers to plant trees in a haphazard manner, the Green Coast project offered a deal that emphasized aftercare. In return for giving their labour for planting, groups of villagers were given small loans to help establish enterprises that would revive their communities – everything from opening cafés to buying new fishing nets or setting up goat farms. Crucially, the deal was that if three-quarters of the trees were thriving after two years, then the loans would be written off.

The incentive worked, says Wetlands International's Indonesian Director, Nyoman Suryadiputra, who masterminded the scheme. Aceh villagers formed groups that planted almost two million seedlings on 1,000 hectares around seventy villages. With better preparation and choice of sites – plus tender loving care from the villagers – more than 80 per cent passed the two-year test, he said. And our visit a decade later suggested that the trees were mostly still thriving. Villagers were

LEFT: The Boxing Day tsunami caused loss of life all along the Aceh coastline, with the destruction worst where mangroves had been removed.

OPPOSITE: A boy plants mangroves along the Aceh coastline. The mangroves will help to provide a buffer against future floods, restore the coastal ecosystem, and revive local fisheries.

FOLLOWING PAGES: New mangroves planted in fish ponds in Aceh in the aftermath of the 2004 tsunami.

proud of them. The dense vegetation was providing visible protection against cyclones, coastal erosion, and any future killer waves. It was also reviving nature. Birds flocked to the cool, new forests. Villagers saw this as a good omen for the revival of their communities.

Some communities were initially reluctant to give up coastal land for planting, particularly when they wanted to restore fish ponds. So, Suryadiputra has worked hard to reconcile the conflicting needs of coastal protection and income generation. Rather than opposing the restoration of fish ponds along the coast, he has instead encouraged villagers to plant mangroves in their ponds and along the dykes between them.

One village that took up the proposal was Krueng Tunong, a fishing community where rescuers found more than a thousand bodies after the tsunami. A local group planted mangroves around 20 hectares of rehabilitated village ponds. It turned out to be a win-win, they told us.

'We get more fish [in the ponds] now that there are mangroves,' local leader Wahab said. 'They grow faster and in greater numbers than when the ponds were bare. I can see the juveniles hiding in the roots of the mangroves. The roots help them avoid predators. We get more crabs, too.'

In Grong-Grong Capa, on Aceh's north coast, they formed a planting group of thirty people, one from each household, who planted 16,000 mangroves around the village. 'They are getting thicker and thicker now. We get lots of cockles around the roots,'

said Nurbaidah, a member of the organizing committee, standing outside her home beside one of the ponds. She picked up a bucket. It contained several hundred large cockles gathered that morning. 'Every day we can pick that amount,' she said. Traders from Banda Aceh, the provincial capital, came to her door and paid ten dollars a bucket. But much as she liked the extra income, the main benefit of the mangroves was 'to protect our homes from the wind and waves,' she said, gathering her children as the sun set behind her. Nothing was more important than that.

Nearby, fisher Mumtadar was waist-deep setting nets in his rented pond. Since he planted mangroves on the banks, the fish have grown bigger and quicker. He harvested 2.4 tonnes a year from his 2 hectares of ponds. Selling at roughly a dollar per kilo, he makes a good living, even after buying larvae and renting the pond, he said. White herons were looking for food in his ponds. But he didn't object. It showed there were rich pickings.

The new post-tsunami landscape of Aceh has attracted visitors from around the world to see how this novel combination of fish ponds and mangroves is being achieved. It is certainly not a return to the days when the coastal strip was covered in forests rich with wild boar, monkeys, and even the occasional tiger. But the combination of natural flood protection and ecological nurturing afforded by the mangroves, coupled with the cash-raising power of the ponds, is a compromise the villagers appreciate.

MANGROVES ARE SALT-TOLERANT trees that grow in shallow tidal waters in tropical regions. They require slow currents, no frost, and plenty of fine silty sediment. Their dense root networks become vital nursery and breeding grounds for marine life, including a large proportion of the world's coastal fish. The coral of Australia's Great Barrier Reef might be barren without the nearby mangroves. Mangroves also yield their own harvests of oysters, which cling to their roots, and crabs and cockles living in the surrounding mud. Along with the algae and filter-feeding animals they harbour, mangroves also clean up pollution, especially nutrients from sewage.

As we have seen, they provide physical as well as ecological services. Mangroves are the best protectors of tropical coastlines. Waves and winds rapidly lose their power as they pass through their dense thickets. They can handle sea-level rise, too. In fact, they thrive on it. The dense roots in a hectare of mangroves can trap up to 20 tonnes of sediment in a year,[297] enough to keep pace with rises up to eight times current rates, says Daniel Alongi of the Australian Institute of Marine Science in Queensland. No sea wall can do that.

But, just when we could use them, mangroves are disappearing fast – at a rate of some 200 square kilometres a year. They are cut for timber or charcoal, and removed for coastal developments such as ports, industrial zones, oil-palm cultivation and, most of all, ponds for raising fish or prawns. Most Caribbean, Indian, West African, and South Pacific mangroves are long gone. Globally, around 150,000 square kilometres remain, says the Global Mangrove Alliance,

Since the tsunami, some of Aceh's coastal communities have restored their inundated fish ponds – but improved coastal protection by planting mangroves in the ponds. They find they get a better fish harvest as a result.

a coalition of NGOs. Most are in South-East Asia, where Indonesia has around a quarter, despite losing an estimated 1,150 square kilometres between 2000 and 2012.[298,299]

Of the seventy known species of mangroves, more than a dozen are thought to be at risk of extinction.[300] But the good news is that the decline can be reversed. In places conservation is happening at scale. Indonesia has established the Sembilang National Park on the east coast of Sumatra to safeguard the largest intact mangrove region in South-East Asia.

Many countries have attempted replanting. A bellwether could be Sri Lanka, a mangrove biodiversity hotspot with twenty-one species. In 2015, the country became the first to declare comprehensive protection of all its surviving mangroves.[301] It recruited 15,000 women in coastal communities to mount patrols to protect the estimated 90 square kilometres of surviving mangroves, and to raise and plant 40 square kilometres of new mangroves in coastal lagoons and abandoned fish ponds. In return, the women get small loans to set up businesses such as bakeries, restaurants, and dressmakers.

Success is far from assured. Sri Lanka has been this way before. A study in 2016 found that, of twenty-three mangrove restoration sites in the country, only three showed more than 50 per cent survival rates, and at nine sites there were no plants at all.[302] Wetlands International has monitored survival rates for mangrove planting in a number of countries. Such failures are all too common. It found that 'seedlings are often planted at unnatural densities and in wrong locations, causing massive die-off'.[295] Typical problems include planting in sand rather than silt, and in locations exposed to scouring tides. 'Tens of millions of euros of public and private conservation funds has gone to waste.'[295]

But restoration can work when the conditions are right. Aceh shows this. And given half a chance, mangroves readily regrow without the need for laborious planting. The waters of many coastal regions still contain a plentiful supply of mangrove seeds, waiting for shallow, silty waters in which to germinate and grow. And rising temperatures are extending their potential range, albeit sometimes at the expense of other coastal wetlands, such as salt marshes.[303]

Guinea-Bissau is a success story. The small West African country has the continent's second largest area of mangroves. This is despite extensive past losses to farmers growing rice. But along the River Cacheu and in the Cantanhez National Forest, hundreds of square kilometres of rice paddy have been abandoned as salt invaded the fields. So now, with assistance from Wetlands International, locals have been stimulating the recovery of hundreds of hectares of former mangrove swamps. Sometimes they have planted new seedlings raised in nurseries. But often all that was required was to break the dykes that surrounded the former rice fields, and let the sea water in. Nature did the rest.[304,305]

The Global Mangrove Alliance wants to expand the global area covered by mangroves by a fifth within a decade. With such initiatives, it is not an impossible task. o

BAY OF BENGAL, SOUTH ASIA
ABSORBING CYCLONES IN INDIA AND BANGLADESH

IN 1999, SUPERCYCLONE KALINGA blew in from the Bay of Bengal across the delta of eastern India's Mahanadi Delta. It claimed more than 10,000 lives.[306] Two-thirds of its victims drowned as a surge of sea water 6 metres high rushed across the densely populated delta. Saudamini Das of the University of Delhi studied what happened at 400 coastal villages near Kalinga's landfall. She says the reason for the devastation was simple. Communities had lost the belt of mangroves that until recently had covered most of the delta's 200-kilometre coastline.[307]

As late as the mid-twentieth century, the mangroves extended inland for an average of 5 kilometres. But by century's end, three-quarters had been cut down and replaced by rice paddies and fish ponds. Das found that nobody died in villages that still had 4 or more kilometres of mangroves; but where protection was less than 3 kilometres, the surge penetrated to the villages behind, and people drowned. The less the protection, the greater the death toll. It was as simple as that.

After Kalinga, India resolved that the loss of life in one of its poorest regions should never be repeated. But it responded with physical rather than natural defences. The government built hundreds of sturdy cyclone shelters, so that next time people at least had somewhere to go. It installed hundreds of new wells and toilet blocks on concrete platforms that would stand clear of future floodwaters. Meanwhile, aid groups helped villages draw up disaster plans, with early-warning sirens, evacuation routes, and family survival kits.[1]

When the next big cyclone hit the delta in October 2013, there were fewer than a hundred deaths.[308] Great news. Many lives had been saved. But the chaos left behind was still considerable. Thousands of straw, timber, and bamboo homes were destroyed, trees uprooted, cars overturned and power lines broken, as the ocean poured across the land. While lives were saved, livelihoods were lost on a huge scale as the rice fields and fish ponds on which most households depended were inundated with salt water. And villagers discovered that, without the mangroves, they had permanently lost land too. The sea had eaten away at their land, bringing the danger from the next cyclone ever closer.

The year after the 2013 cyclone, during a tour of the delta, we visited the coastal villages of Tandahar and Keutajanga, which sit either side of a creek carrying water from the River Mahanadi to the ocean. 'The sea is coming ever closer,' said Pramod Swain, Secretary of Tandahar's disaster committee. 'It has come in about 1.5 kilometres. Our fathers' land is now under the water.' Over the river in Keutajanga they said that the 2013 cyclone had broken their sea wall and permanently flooded 40 hectares of fields on the edge of the village.

Some villagers were fatalistic about the future. One day soon, their villages would be washed away. But not all. In Keutajanga, they wanted to show us the mangroves they had planted between the shore and their village to reduce erosion and absorb future cyclone surges. In Tandahar, village women had formed a committee to look after their remaining trees and plant more. They fined anyone who cut branches or let livestock into the new woodlands.

The villagers also told a more complicated story of life on a low-lying delta exposed to cyclones. A story that suggests restoring the mangroves is not all that is required to secure their livelihoods from floods. During the 2013 cyclone, some 5,000 square kilometres of delta farmland had been flooded. But much of the floodwater, they said, had not come from the ocean. Days after the ocean retreated, as they got ready to put away their cyclone survival packs, they had been hit by a second wave of flooding. This time it came from inland. The cyclone, it turned out, had blitzed the entire catchment of the River Mahanadi with heavy rain. And while emergency services were concentrating on repelling the threat from the ocean, that water had come down the river, bursting its banks and spreading out across the delta.[309]

In the past, this water would have collected in low-lying areas within the delta, and eventually evaporated. But since 1975, nearly 30 per cent of the delta's huge network of ponds and wetlands have been drained to make way for fields, fish ponds, and industrial zones.[310]

Fragile homes made of wood and palm fronds are destroyed when supercyclones rip across the Mahanadi Delta.

FOLLOWING PAGES:
Tidal flooding has long been a feature of the Sundarbans, the world's largest stand of mangroves, which occupies the delta region at the head of the Bay of Bengal. But flood defences and land reclamation in some areas are concentrating incoming waters, creating higher tides in neighbouring areas.

Without the wet places, the water had nowhere to go. Rivers rose. Fields flooded. Dykes intended to keep floodwaters away from villages were overwhelmed. Schools and homes were inundated. Once water got inside the banks, it was very hard to get rid of. When it did finally go, it left behind fields smothered in sand.

The lessons here are nuanced. When cyclones come, shelters do save lives. More mangroves on the delta coastline would save more, while also slowing erosion and protecting against rising tides. All that is essential. But only a wider restoration of wetlands within this and other deltas can protect people from flooding that comes from inland. In the old days, the landscape was designed to accommodate floods rather than repel them. Some of that resilience needs to be recovered, says Wetlands International's South Asia head, Ritesh Kumar. 'Living with floods is the key. We can't go back, but we do need to introduce elements of the old regime. We need to rethink embankments,

because they restrict the ability of the floods to spread across the delta.' When floods come, they need somewhere to go.

THREE HUNDRED KILOMETRES north-east of the Mahanadi Delta lies an even larger delta, through which the waters of the Ganges, Brahmaputra and Padma rivers all reach the Bay of Bengal. It straddles the border between India and Bangladesh. Close to its centre lies the ramshackle river port of Khulna. The third largest city in Bangladesh, one of the world's poorest and most populous nations, Khulna is among the most flood-prone urban areas on Earth. The city was built around hundreds of ponds, linked by canals. They once called it the Venice of Bengal.[311] Now most of the ponds have been drained for real estate – except that the drains don't drain when tidal waters flood through the delta, up the drains and into the city's streets.

Khulna is literally disappearing beneath the water. Despite being 125 kilometres inland, the homes of a tenth of its million-plus people are flooded most years. It may only have a decade left before life here becomes impossible.[312] What can be done? Can nature's flood defences be harnessed to meet the new threats? And are we sure that our current flood-prevention strategies are not making things worse?

Almost half of Bangladesh is inside the vast three-river delta. The incessant delta processes of erosion and sedimentation of silt and sand are natural. But that is small consolation to the thirty million people living here, many on sand banks that are at constant risk of being eroded or engulfed by the surrounding waters. Tragedy is routine. Floods that would be front-page news in other countries, with dozens of villages disappearing beneath the water, barely register in Bangladesh.

Traditionally, the Bengali delta people adapted to this environment by constantly being ready to abandon their temporary homes and move their belongings somewhere else. For as one sand bank is submerged, another most likely forms. The River Padma has eroded more than 660 square kilometres of land along its banks in the past fifty years, but created a similar amount.[313] In many areas of the delta today, however, more land is being lost than gained. A fast-increasing delta population is being left with ever-fewer options for relocation.

There are two conventional explanations for this loss of delta land. One is that climate change is raising sea levels generally, leaving the delta exposed to both insidiously rising waters and the more sudden impact of tidal surges unleashed by cyclones. That must be true. But it is only part of the story. For, while sea levels in the Bay of Bengal are rising by 3 millimetres a year, high tides in Khulna and much of the rest of the delta are rising up to six times faster.

The last of Mousini Island in the Sundarbans. Three-quarters of it has been engulfed by rising waters.

Some of this extra rise may be explained by the loss of mangroves. But despite their widespread conversion to rice cultivation and prawn ponds, mangroves still cover much of the delta. The area known as the Sundarbans is reckoned to be the largest mangrove swamp in the world. It covers 10,000 square kilometres, and remains home to Bengal tigers, river dolphins, and estuarine crocodiles. So, mangrove loss may not be the whole story either. Some hydrologists say there is a third element in this disaster – one with important implications for many delta regions around the world. It seems that Bangladesh's efforts to protect delta areas from ocean floods are making things worse, not better, especially for places further inland such as Khulna.

It happened like this. The Government of Bangladesh has for half a century been building thousands of kilometres of earth embankments along the numerous waterways of the delta. This work for a long time soaked up much of the international aid given to the country. During the 1960s and 1970s, USAID paid for 1,600 kilometres of embankments around Khulna alone. Elsewhere, Dutch engineers built polders. The whole of Bangladesh was going to be flood-proofed. In 1990, one of

LEFT: Intact mangroves, such as these in the Sundarbans, hold back high tidal surges and buffer against typhoons. Their loss is exposing one of the most vulnerable coastlines in the world.

OPPOSITE: Women head for the beach to sort fish caught off the beach of the Mahanadi delta in India. The ocean is a source of natural wealth for the delta's people, but also a threat when tropical cyclones blow in.

us (FP) met a Dutch engineer in the country's capital, Dhaka, who declared with confidence that 'in future Bangladesh will look more like our country. The only question is when it will happen.'

When properly maintained, such embankments do of course provide some protection for the people and property inside them. But they have a perverse effect outside. By keeping some land safe, they constrict and funnel tides in the surrounding waters, raising water levels and pushing the water further inland.[314] Just as levees on rivers amplify river floods downstream, so embankments on deltas push tides inland, making the lethal surges generated by tropical storms even higher than before.

Once, tidal surges were partially absorbed and dissipated by the mangroves. But the funnelling effect of the embankments does the opposite. A study of tidal gauge data by Shahjahan Mondal of Bangladesh University of Engineering and Technology found that

in south-west Bangladesh, the 'maximum tidal high-water level is increasing at a rate of 7–18 millimetres a year',[315] which is an increase of between two and six times faster than at the coast. The highest figure was recorded in Khulna. An analysis for the World Bank by British geographer John Pethick pinned the blame for this on the funnelling effect.[316] It warned that in combination with an anticipated 1-metre global rise in sea levels, it could deliver a 2.5-metre rise in high tides in places such as Khulna by the end of the century. The city and hundreds of villages around it would have drowned long before high-tide levels rose as high as that.

Pethick's findings raised concerns about the advisability of a $400-million upgrade of 600 kilometres of coastal embankments south of Khulna funded by Pethick's commissioners at the World Bank. He says his findings were accepted by the bank's staff in Washington DC. Most experts in coastal processes agree with Pethick that embankments in the delta will probably expose at least as many people to floods as they protect. Marcel Stive, a coastal engineer at Delft University of Technology in the Netherlands, told us, 'When one embanks estuaries … the tidal range increases.' According to Julian Orford of Queen's University Belfast, who collaborated with Pethick in a published version of the World Bank study, 'making embankments worsens the flooding problem.'

But the Bangladeshi Government remains keen on hard flood defences, and opposes restoring mangroves or other natural flood defences, because that would mean 'abandoning' some of the delta lands currently occupied by its citizens. The embankment project goes ahead. 'People are very reluctant to take notice of data that is inconvenient,' says Orford.[315]

What should be done? Coastal engineers say the findings of amplified tidal peaks provide more ammunition for the call to replace so-called 'hard engineering' wherever possible with 'soft' flood defences that work with natural forces rather than confronting them. Building with nature, in other words. Pethick says Bangladesh may still need some embankments. But they should be set back much further than before. That would give more room for tidal surges to spread horizontally – so lowering their height. That set-back strategy should be accompanied by restoring the mangroves and other ecosystems that can provide natural protection against incoming storms. As Colin Thorne of the University of Nottingham put it, 'Ecosystems evolve and adapt to change; man-made structures do not. The best hurricane protection is three kilometres of mangroves.' o

WADDEN SEA, NETHERLANDS
BREAKING DYKES TO STOP THE OCEAN

REDSHANKS NESTING ON THE BANKS of the creek sang in unison as we approached. It was a warning. Humans were invading their domain. But Chris Bakker could claim his own right to be there. But for his NGO, It Fryske Gea, these marshes would still be poldered, with the sea kept out to maintain cattle pasture. Our walk came almost twenty years after an excavator more accustomed to raising dykes to protect the Netherlands from floods, had instead taken a bite out of a polder bank at Noard-Fryslân Bûtendyks, north of the Friesland town of Leeuwarden. From that day on, more than 120 hectares of former dry land re-joined the Wadden Sea, a muddy intertidal zone of salt marshes and mud banks that is the largest continuous area of tidal flats in the temperate world.

The Wadden Sea is an often-forgotten jewel in the crown of European conservation. It extends almost the length of the Netherlands, and across the north of Germany as far as Denmark. It is a vital spawning ground for the fish of northern Europe. Some 90 per cent of the North Sea's herrings begin their life here, for instance. In summer it is visited by around ten million migrating birds. Yet, marvellous though it is, the Wadden Sea is a shadow of its former self. Some 80 per cent of its salt marshes have been reclaimed since 1600, by more than 200 kilometres of dykes.[317]

As late as the 1980s, the Dutch saw this land reclamation as work still in progress. Now the narrative has changed. The growing cost of maintaining the sea walls in the face of rising sea levels is regarded as prohibitive. A nation largely built by reclaiming marshland from the Wadden Sea and the Rhine Delta has decided enough is enough. Just as the Dutch are giving their rivers room, so they are giving land back to the sea. Engineers who once exported their hard engineering around the world have a new creed along the shore as well as in the Rhine Delta: building with nature. It's good news for redshanks.

Polder is a Dutch word, because they invented the technology of building barriers around shallow waters and draining the area inside to create new land. Now they are inventing depoldering, the art of breaking sea walls to turn fields or cattle pastures back into salt marshes. The aim is not to give in to the ocean. It is to encourage natural processes that will capture sediment to protect the coastline and raise it naturally with the tides. To harness what Jantsje van Loon-Steensma of Wageningen University calls 'the positive feedback between salt-marsh vegetation and sedimentation' that effectively turns salt marsh plants into eco-engineers that change their own environment.[316] Salt marshes in temperate lands do this almost as effectively as mangroves in the tropics.

At Noard-Fryslân Bûtendyks, the breaking of the sea wall has expanded the old salt marsh by 120 hectares, just over a square kilometre. It is a small step. Conservationists such as Bakker would like to go much further. They have a detailed blueprint for depoldering 10 square kilometres on this section of coast. But the ultimate aim, they say, is to do ten times more, expanding salt marshes here from the current 33 square kilometres to closer to the former 140 square kilometres. But that depends on gaining public consent, which may not be easy says local politician Johannes Kramer of the Friesland Provincial Executive. 'The forefathers of some of the local people created the land with their bare hands,' he points out.[318] Handing it back to the ocean 'stirs up strong feelings'.

Breaking the dyke at Noard-Fryslân Bûtendyks did not cause a sudden inrush of sea water. Instead, the sea gradually found its way up creeks and has slowly infiltrated the ground. It floods the former

Almost 60 per cent of the mudflats in the Bohai Sea, the innermost gulf of the Yellow Sea, have been reclaimed.

FOLLOWING PAGES:

At Noard-Fryslân Bûtendyks, north of the Friesian town of Leeuwarden in the Netherlands, grazing land along the shore is being turned back into salt marsh, as a means to protect fields inland.

polder about twelve times a year. The purpose is to allow the salt marsh to regrow so that it absorbs rising tides and protects the sea defences further inland. The marsh is certainly taking hold. 'Within a year we had found all twenty-four of the local salt-marsh plant species, including glasswort, sea plantain, and common seablite,' said Bakker. Some birdlife has come back, too. Barnacled geese have increased, for instance. But overall, birders have been disappointed. The remnants of the old polders so far remain better habitat for many species, including avocets, oystercatchers, terns, and many meadow birds.

The new marsh is still a managed landscape.[319] Bakker and his colleagues have agreements with local farmers who put their cattle out to graze on it in May or June – after flocks of grazing geese have had their fill and departed. Conservationists, he said, are learning as they go along how to reconcile grazing and marsh ecology. There are, it turns out, win-wins. Horses and cattle seem to be good for marsh biodiversity, for instance. Their grazing prevents the spread of cordgrass, which can overwhelm other salt-marsh plants. But the animals also trample birds' nests, and restrict the height of vegetation that captures silt washed onto the marsh by the tides.

Even so, the primary purpose of the exercise is being achieved. The first decade of rewetting at Noard-Fryslân Bûtendyks saw marsh plants raise the land surface by an average of 7 centimetres. That was more than twice the sea-level rise over the same period. The new marsh is making up for lost time, said Bakker. A new shoreline more resilient to the power of waves and rising tides is forming.

THE EVIDENCE IS GROWING here on the Wadden Sea and elsewhere of the power of salt marshes to protect shorelines from erosion and rising sea levels. A barrier of salt marshes as little as 10 metres wide can halve wave heights, concluded Christine Shepard of the University of California at Santa Cruz, after a detailed analysis of seventy-five different studies, from the California coast to the English North Sea and the Yellow Sea in China.[320] Five hundred metres of marsh typically reduces wave heights by 90 per cent. And they do more. They raise the land. Some studies suggest that, by capturing sediment, salt marshes should be able to keep pace with sea-level rise of as much as four times the current rate. Much will depend on local conditions, however.[321]

The issue could be critical in south-east England, where coastal managers face a particularly severe crisis because rising tides are combining with land that has been sinking since the end of the last Ice Age. Coastal engineers anticipate that subsidence along the Thames Estuary near London will double the impact of sea-level rise.[322] After a tidal surge in the estuary in 1953 drowned more than 300 Britons, they fought back with concrete and rip-rap, culminating in the construction of the Thames Barrier to protect London. But their efforts have had a perverse effect on some of nature's coastal defences. Twenty per cent of the salt marshes along the outer Thames Estuary in Essex and Kent disappeared in the last quarter of the twentieth century, often as a result of being squeezed by ill-sited sea walls.[323]

Recognizing their mistake, engineers are now removing sea walls to allow salt marshes to take over once again. They began at Wallasea Island at the mouth of the River Roach in Essex. The island was first reclaimed from the sea by Dutch engineers 500 years ago. But with the salt marshes that once surrounded the island gone, and sea walls crumbling, in 2015 the island's owners brought in excavators to breach the walls and allow the sea to wash over former fields and restore the old ecology. So far, salt marshes, mudflats, and lagoons have begun to reform across around 600 hectares of the island.[324] It is a small step. Four centuries ago, there were 300 square kilometres of tidal salt marsh along the Essex coast. Today only around a tenth remain. But it may be the first of many such projects.

All this is part of a growing rethink about how to defend the world's low-lying shores against rising sea levels. A key conclusion is that natural defences are not only cheaper but also work better. This is bad news for much of Europe where, in recent decades, wetlands on the Atlantic, North Sea, and Mediterranean coasts have been lost on a large scale. But it is good news for North America where, perhaps because it is less densely populated, two-thirds of the coastline is still occupied by wetlands and other natural features. Environmental economists a decade ago put the total value of these coastal wetlands

Seafood for sale at a market in Dalian, a town on the shore of the Bohai Sea, whose mudflats have been largely commandeered for industrial development, including for petrochemicals. Seafood sales have been badly damaged by oil spills.

in storm protection alone at $23 billion. Every hectare was potentially worth $33,000.[325]

Such economic assessments are gaining traction with policy-makers. They can see the monetary value in protecting natural defences, and increasingly see potential in replacing what has been lost. On the estuary of the Delaware River, the second largest estuary on the eastern coast of the US, salt marshes had until recently been disappearing at a rate of more than 2 hectares a week. To reverse that loss, the authorities dumped bundles of coconut fibre onto the shore. These 'biologs' acted as surrogate salt marshes. They dissipated wave energy and captured sediment to kick start the regeneration of marshes. The effects were immediate. In one pilot area, healthy salt-marsh plants established themselves within a year. Mussels were so keen to get going that they attached themselves to the biologs and began filtering polluted water. A wetland was being reborn.[326,327]

Anything America can do, China, these days, reckons it can do better. Nowhere in the world has coastal infrastructure development been as rapid as in China. In recent decades, marshes and mudflats have been turned into ports and drained for farming and urban sprawl on a staggering scale. China's concrete sea walls are now longer than the Great Wall.[328] Half the country's coastal wetlands have already been lost, the Chinese Academy of Sciences concluded in 2015.

If anything, the intertidal zone has fared even worse. There are sand mines on the beaches, and wind farms and oil rigs sprouting across mudflats and salt marshes. Almost 60 per cent of the mudflats in the Bohai Sea, the innermost gulf of the Yellow Sea, have been reclaimed. Those flats were a vital staging post on the East Asian–Australasian Flyway, which is used by some fifty million birds a year. That role is under extreme threat, say ornithologists. Numbers of the critically endangered spoon-billed sandpiper have fallen by 90 per cent in forty years.

But China is turning the corner. Beijing has announced that there should be no more coastal development on the shores of the Yellow Sea. And in 2018 it signed an agreement with Wetlands International to collaborate on helping secure the remaining wetlands, in particular those used by migrating birds, and to roll back some of what has been lost.[329]

China's actions need to be matched elsewhere in the region, however. The Yellow Sea's coastal wetlands are also under siege in South Korea, where a decade ago the government built the world's largest sea wall, extending 34 kilometres around the Saemangeum Estuary. Until 2006, the estuary was a maze of mudflats and silty water visited by hundreds of thousands of migrating birds. Local women harvested clams and octopuses.[330] But since the wall was completed, the shellfish have expired, and the birds that once came to eat them, including the spoon-billed sandpipers, have gone elsewhere.

The Government in Seoul built the wall with the promise of turning the 3,000 square kilometres of dried estuary into a 'green city'. More than a decade on, there is no sign of any city, green or not. Instead, there is a plan for an energy park to generate wind and solar power.[331] Environmentalists have called for the government to admit its mistake, open the sluice gates in the wall and allow the water back in. Wind turbines and solar panels do not need a lot of land. They could be green-energy islands in a restored estuary. ○

FLORIDA EVERGLADES, USA
START OF A BLUE CARBON REVOLUTION

They call it the most expensive wetlands restoration project in history – to return the Everglades of southern Florida to something like its former glory. The Everglades once covered 10,000 square kilometres. At its heart was the largest freshwater marsh in the US, a vast expanse of waterlogged sawgrass that for most people is the defining image of the Everglades. This 'river of grass' had over time created a layer of peat on average 2 metres deep. The marsh exited into the ocean through the largest mangrove swamp in the continental United States, covering around 1,500 square kilometres, as well as cypress swamps and sea grasses.

But after decades of urban encroachment, pumping of freshwater to fill city taps, large-scale drainage to create sugar-cane farms and dyke-building to prevent floods, the wetland was reduced to less than half its former size by the close of the twentieth century. The sawgrass was a much-diminished and artificial managed ecosystem. Lake Okeechobee, the body of water at the centre of the marsh, was a sump for farm drainage. Thanks to oxidation, the carbon content of the peat had reduced by four-fifths.[332]

Now the turnaround has begun. Federal legislation passed in 2000 signed off $8 billion to be spent over forty years to reverse environmental damage and rewet some 4,000 square kilometres of peatlands.[333] Dykes would be torn down. Rainwater would be captured and diverted into the wetland. The river of grass would be restored. The

LEFT: Large swathes of the Florida Everglades have been drained, often to create prime real estate. To ensure coastal views, mangroves have been widely chopped down.

OPPOSITE: Islands of mangroves thread through the Everglades National Park in Florida, one of the world's largest subtropical wetlands.

FOLLOWING PAGES:

Wildlife still thrives in the Everglades when it is given the chance. This eastern diamondback rattlesnake is endemic to the region.

mangroves, which hold a further 40 million tonnes of carbon, would be protected and expanded, creating habitat for iconic species such as the Everglades' manatees, alligators, sea turtles, and Florida panthers.

Not all is going to plan. The cost of the extensive engineering works to put fresh water back into the marshes has doubled, while funding for the work by the Corps of Engineers has been cut back. Some scientists say sea-level rise is filling the marshes with salt water faster than the Corps can restore freshwater inflows. The National Academy of Sciences estimates it could now take another hundred years to finish the job[334] – if rising sea levels don't make the works redundant. But optimists still see the restoration as a centrepiece of both ecological restoration in the US and America's future efforts to keep carbon out of the atmosphere. The carbon in the mangroves alone is valued by environmental economists at more than $2 billion.[335] The Everglades,

says the Academy, could become a global pathfinder for conserving and restoring 'blue carbon' – the carbon in coastal wetlands.

MANY HEALTHY WETLANDS, as we have seen, capture and store huge amounts of carbon. But like forests, if the wetlands are lost, that carbon surges into the atmosphere adding to climate change. We have seen growing concern among climate scientists about peat bogs from Western Siberia to the tropics, and how their demise may be releasing around 2 billion tonnes of carbon dioxide into the air annually.[336]

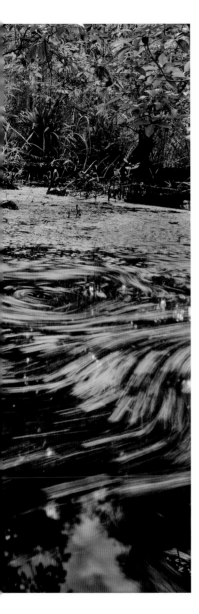

A cypress swamp filled with duckweed in the Big Cypress Reservation of the Seminole Tribe in the Florida Everglades.

There are equal concerns about loss of carbon in the world's coastal wetlands – 'blue carbon'. Mangroves, salt marshes and sea-grass beds are immensely productive biological systems. Their photosynthesis is in overdrive, and its main feedstock is carbon dioxide grabbed from the air. A hectare of mangroves typically takes up 6–8 tonnes of carbon dioxide per year, according to the UN's Intergovernmental Panel on Climate Change.[337] Globally that adds up to more than 30 million tonnes of carbon a year – roughly the same as all the world's tropical rainforests.

The world's salt marshes are even better. They absorb almost three times as much carbon from the air as mangroves – up to 87 million tonnes a year.[338] Sea grasses, marine flowering plants below the tideline, perform a similar role and may sequester as much as 100 million tonnes. From the mangroves of Belize to the salt marshes of New England, carbon deposits have been found in coastal sediments that are as thick as peat bogs and thousands of years old.[339] Coastal wetland ecosystems currently hold up to 25 billion tonnes of carbon, according to an assessment carried out in 2018 by the US National Academy of Sciences.[340]

That is all good news, except that these carbon stores are diminishing. Nobody can be sure of the numbers, but an estimated 25 per cent of salt marshes and 30 per cent of mangroves have disappeared as a result of coastal development in the past sixty years. The resulting carbon dioxide emissions to the atmosphere are probably around 500 million tonnes annually, and perhaps double that.[341,342]

Some of the carbon from eroding coastal wetlands may float off in coastal sediment and help build wetlands elsewhere. Nature has always done that. Global warming may even encourage the growth of mangroves in new places.[335] But the chances of recycling lost blue carbon may be limited because coastal wetlands are running out of room. They are being squeezed between rising sea levels and the concrete and rip-rap of coastal developments.

The loss need not continue, however. Indeed, for the sake of the planet's climate, they must not. New policies to protect and revive coastal wetlands could turn them back into absorbers of blue carbon, says Daniel Murdiyarso at the Center for International Forestry Research, which is based in Indonesia. The world's richest blue-carbon stores are in Indonesia's mangroves and sea grasses. Fringing the country's 100,000 kilometres of coastline, they contain around 3.5 billion tonnes of carbon.[343]

Murdiyarso says that a moratorium on the further loss of mangroves in Indonesia could, with one leap, allow the country to meet its climate targets made in Paris in 2015. It would also build the country's resilience to coastal storms, and improve the livelihoods of people living on its coasts. In 2018, the country agreed a National Mangrove Ecosystem Management Strategy to help realize this. Other countries could follow suit. As part of the Paris Agreement, twenty-seven other nations joined Indonesia by including mangrove restoration as part of their national plans for limiting greenhouse-gas emissions.

Countries in temperate lands could take similar steps to preserve and restore blue carbon in salt marshes. British researchers in 2019 completed a study of nine sites in eastern England where salt marshes had grown back, usually after their destruction by storms. The results were most encouraging. Within a century they had captured an average of 74 tonnes of carbon per hectare, slightly more than held in nearby undisturbed marshes.[344]

As we have seen, countries such as the UK and the Netherlands already set aside coastal areas for recovering salt marshes, as a way to head off rising sea levels caused by global warming. The same, says the US National Academy of Sciences, could be done by many countries, including the US, to grow their natural carbon stocks and head off global warming itself. The Academy reckons the US has an estimated 13,000 square kilometres of tidal wetlands and sea-grass beds that could be restored.[345] Worldwide, it calculates that 5 billion tonnes of extra carbon dioxide could be stored in coastal wetlands by 2100. That is not on its own a game-changer for our climate. The figure is almost the same as the US's annual emissions from industrial activities. But it would help.

Of course, we shouldn't need a climate incentive to restore coastal wetlands. This book has laid out in detail the importance of reviving wetlands for wildlife and conservation, for flood prevention, and to assist the poor and landless. But, in a world concerned by climate change, the carbon-capturing potential is an additional incentive. o

JAVA, INDONESIA
'GOD WILLING, WE PLAN TO STAY'

To reach Timbulsloko, a village with a population of around 3,500, we drove for 5 kilometres along a narrow causeway. All the way, a single line of houses was strung out on either side of the road. But behind them, rather than the expected fields, there was a lot of water, punctuated by half-submerged fences and the remnants of dykes, with occasional newly planted mangroves on top. Two decades before, this had all been land; but since then the ocean had encroached. Richer residents along the road were rebuilding their houses on new, higher foundations to keep above the rising tides. Other houses were inundated, abandoned, or marooned on tiny islands that could only be reached by rickety walkways. The village cemetery was being washed away. Locals said lapping waters occasionally floated decomposing bodies into their living rooms.[346]

At the far end of the causeway, in the village hall, village elders discussed the plight of their community, and remembered the old days before the water came. The village was on a delta that had long been a prosperous rice-growing area, famous for its fertile soils. It had been protected from the ocean by a residual belt of mangroves. 'I grew up in the 1960s when the sea was more than a kilometre away,' said Slamet, a fisherman. 'Then the flooding began.'

Beside him, village activist Mat Sairi admitted they had made mistakes back then. Like almost every coastal community in the area,

LEFT: Parts of Semarang, a booming city on the northern shore of Indonesia's populous island of Java, are regularly under water as the land sinks. But the city's reliance on underground water is partly responsible for the sinking,

OPPOSITE: Sinking land and the conversion of mangroves to fish ponds has created a crisis along Java's northern shore. The ocean has invaded several kilometres inland, flooding rice fields and obliterating whole villages. Things are worst here in the district of Demak.

they had wanted to make money by raising prawns and milkfish in ponds. Traders were paying good prices, and salt water was invading the rice fields. So, the villagers made room for the ponds, first by converting their rice fields, and then by cutting down the mangroves along the shore. 'Our parents warned us that we should protect the mangroves,' he remembered. 'They said the mangroves provided many benefits, like the oysters, crabs, and fish among their roots, as well as protection of the coastline. But our people wanted to make money and feed their families.'

By the 1990s, with the mangroves gone, the sea began to wash away the dykes that surrounded their fish ponds. By 2013, the sea had penetrated inland for more than a kilometre.[347] They had lost twenty-five rows of fish ponds. Now the waters lap at their houses. Other nearby villages have been washed away. They feared they might be next.

Indonesia has more low-lying coastal regions than any other country in the world. Once the sea was kept at bay in most places by a kilometre or more of mangroves. No country had so many. But as the coastal lowlands became more densely populated, the mangroves were replaced by rice fields, and then by fish ponds and other economic activities such as ports and industrial areas. Indonesia's most heavily populated and economically developed island, Java, has lost more than 70 per cent of its mangroves.[348] This opened up the coast to rapid erosion. And nowhere has suffered more than Timbulsloko and a clutch of other

villages in Demak, a low-lying district on the north shore of Java.

Demak – a name that comes from the Arabic word *dhima*, meaning swamp – was in past centuries a prosperous sultanate with rich rice paddies sheltered behind the mangroves. Its port dominated trade across what is today Indonesia. That dominance is long gone. Though retaining its grand mosque, the town has ceded economic power to other Javanese cities. Now its hinterland of rice paddy is disappearing altogether from the map.

In places, the Demak coastline has retreated by more than 3 kilometres. Two communities, Senik and Tambaksari, have been

Villagers in Demak are fighting back against the rising tides. Here, at a field school that Wetlands International helped establish, they are learning innovative techniques to stabilize the coastline and restore mangroves.

FOLLOWING PAGES:
All along the Demak shoreline, village groups are erecting brushwood barriers aimed at catching and holding silt to create a base. Mangrove seeds floating in the water will be trapped, germinate in the silt and re-establish, stabilizing the coastline.

submerged, and their people evacuated. Twelve others have been badly impacted. At a meeting in Demak's district offices, officials said 19 square kilometres of land in the district has been lost. In 2017 alone, more than 500 people lost their homes. A thousand fish ponds covering 500 hectares have been engulfed by the waves.

Our tour of Demak villages revealed the human stories behind this tragedy. The village of Bedono was, like Timbulsloko, strung out along a raised road out into what is now sea. Driving along the causeway, we constantly had to wait behind trucks delivering sand and bricks to raise up waterlogged houses. The village community hall, nicknamed Bedono Island, was built on stilts. Mudskippers and crabs scurried beneath. 'Before, we had a good life with rice and fish,' said the young village head, Agus Salim. 'Now we only have a memory of agricultural land.'

In Tugu village, which is 3 kilometres inland, Muhammad Masrur gave us tea in his front room. 'Tugu is higher than the other villages, and we have no coastline,' he said. They thought they were safe from the ocean, until the water in their irrigation ditches began to get saltier. It got so bad after 2010 that they began converting their rice fields into fish ponds. 'Then in 2017, my house flooded,' Masrur said, showing pictures of his living room knee deep in water. 'The flood lasted five days. It was salty water, and it came from Timbulsloko. I fear our village may not be here in fifteen years.'

Wedung, a large village on the delta of the River Wulan about 15 kilometres north of Timbulsloko, is away from the coast and high on a river bank. The houses of its 12,000 residents are dry for now. But many of their fish ponds nearer the coast have succumbed to incoming tides. 'We've lost 500 metres to the sea in the last ten years,' said Maskur, a teacher and member of the village committee, as our boat headed out of what is now the river's mouth into a bay that is unmarked on any maps. In places, we could see the submerged remains of banks that had once surrounded fish ponds. But in place of the ponds, there were nets set to catch wild crabs.

'I bought 10 hectares of ponds here in 2004, but three years later they were swept away,' said Maskur's deputy, Nor Khamed. 'If God wants it to happen, it will,' he added with a shrug as a call to prayers rang out from the village mosque. 'You have to accept it. But I don't want my neighbours to experience the same thing, so I want to stop the sea coming in.'

We heard the same mixture of fatalism and determination in each of the villages. 'We are not leaving. This is our home and, God willing, we plan to stay,' said Slamet in Timbulsloko. People had recognized the folly of removing the mangroves. Now they were banding together to

remake the coastline. And rather than calling on the authorities to put up concrete sea walls, they hoped to restore their lost land in a novel way. They were erecting long permeable brushwood structures in the waters that once marked the shoreline of their community.

We took a boat out from Timbulsloko to see some. The brushwood structures are substantial. Each is 175 metres long and rises more than a metre above the waves. They are made up of two lines of vertical bamboo poles hammered 2 metres into the sea bed, with the gap between them filled by a mass of horizontal brushwood held in place by netting. The structures were not intended to hold back the tides.

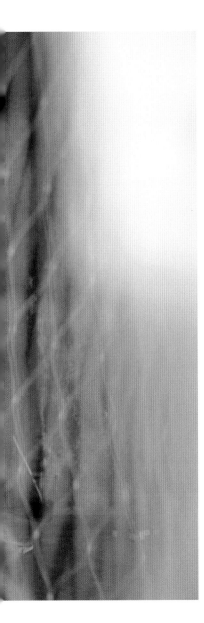

New nets being woven for fish ponds in Demak.

The idea was that the brushwood would slow the waves as they passed through. This would cause the tiny particles of sediment carried in the water to drop down and collect on the landward side of the structures. This new sediment should provide a stable base where mangrove seeds floating in the water could germinate and grow.

In essence, they are a new and enlarged version of an old traditional Dutch technique for catching sediment to stimulate the growth of salt marshes along the Wadden Sea. Here, the Dutch willows are replaced by Indonesian bamboo.[349] If all goes to plan, the structures will eventually recreate a 'green belt' of mangroves that will continue to trap sediment, thus ending erosion and restoring the coastline. The growing mudflats should also provide food for water birds such as egrets, herons, and the endangered milky stork, said Yus Rusila Noor, Wetlands International's Indonesian biodiversity specialist.

Timbulsloko was the first community to erect them. By the time of our visit in mid-2018, there were eight off Timbulsloko alone. Meanwhile, nine other communities had joined in. All told, villagers had erected them along more than 4 kilometres of Demak's coastline.[350] Construction had been laborious. Each structure required the work of twenty-five people for four weeks. Some had suffered damage. After clams known as shipworms began to eat the bamboo poles, some were replaced by PVC pipes filled with concrete. Others were repaired after storm damage in late 2017. In early 2019, some 1.8 kilometres of community-built structures remained in place.[349]

Were they working? Reaching the first brushwood structure, Mat Sairi put a paddle into the water on either side of the barrier to check the water depth. The brushwood had only been in place for eight months, but the sea bed was already 15 centimetres higher on the landward side. It was clear evidence that the structure was capturing sediment behind it, he said, though it might be a while before new mangroves appeared. The aim is to prevent erosion that often reaches 100 metres a year along the Demak coastline, and ultimately to prevent the anticipated loss of 60 square kilometres of ponds and other land, predicted to occur by 2100.

The erecting of the structures is the most high-profile element of an initiative known as Building with Nature Indonesia. It is a joint initiative by Wetlands International, the Indonesian Ministry of Marine Affairs and Fisheries and the Ministry of Public Works and Housing, together with local communities and other partners. The villagers are not paid to erect the structures. Instead, like the mangrove planters in Aceh, their groups are offered a deal. In return for their labour – erecting and maintaining the barriers, and working on other green initiatives such as reviving mangroves along river banks and shorelines – village groups are offered field training in coastal management, plus a loan for local sustainable development projects.³⁵¹ Ten community groups in nine villages have shared just over $300,000 in such loans. In 2018, ownership of the structures was formally transferred to the communities. If the structures are kept repaired and the growing mangroves are protected, then the loans will be written off.

Villagers have used the loans to invest in a wide range of economic activities. More than 400 hectares of fish ponds now benefit from

LEFT: Milky storks are mangrove specialists and are returning as mangrove habitat is restored.

OPPOSITE: Several villages in Demak have constructed boardwalks through their new mangroves as a tourist attraction.

new practices learned during the field courses, while 70 hectares of degraded ponds have been earmarked for mangrove revival.³⁴⁹ In Wedung, women have set up a business to make pellets out of dead fish and cassava flour to use as an organic fertilizer in their surviving fish ponds. One woman elder, Siti, took us to see a pond where they had tried out the pellets in combination with aquaculture techniques they had learned in their field courses. 'With chemical fertilizers, it takes four months to get a harvest. With the pellets it is only two months, and they are cheaper.'

The prawn harvest was under way. We watched as Siti supervised three men wading through the pond picking out prawns by hand and throwing them into a boat pulled by one of the women on the bank. Fifty prawns weigh a kilogramme, Siti told us. At local markets, a kilogramme fetches the equivalent of five dollars. Since learning the new techniques, they have tripled their incomes, putting a proportion of the profits into a community savings fund.

Other villagers have branched out into tourism. Constructing boardwalks through newly planted mangroves is a favourite. One boardwalk, on the river bank near Wedung, has a bamboo mock-up of a miniature Eiffel Tower as an elaborate gateway. Maskur said local tourists come at weekends. To beat off rivals, they plan to include a boat tour. At Bedono, their boardwalk meanders through thick mature mangroves and has a tower with views over the top of the trees

towards the ocean. It is dotted with signs on colourful boards, offering messages to visitors such as Protect Nature, Be Happy and, more ominously perhaps, Pray.

Villagers are still praying that their efforts to regenerate mangroves will work. How long will it be before there are naturally seeded mangroves growing in the silt behind their structures? 'There is no fixed time. It depends on the waves and sediment,' said Apri Susanto Astra of Wetlands International. Four years might be a minimum. 'But if the natural mangrove growth does not occur, it can be helped by spreading seeds on the new sediment,' he said.

THE VILLAGERS APPEARED ORGANIZED, disciplined, and eager to maintain their structures. They know the work could be the only lifeline for their communities. And the district authorities have an agreed plan to use mangrove restoration as the prime tool in restoring 20 kilometres of Demak coastline, reducing coastal hazards and making aquaculture sustainable.

But there is a problem. Mangrove loss, it turns out, is not the only cause of the marine invasion. During our meeting in Timbulsloko, Sifatul Khoiriyah, a teacher in the village and member of the community group, pointed along the shore towards a rapidly growing industrial zone 6 kilometres away. It was on drained low-lying marshland in the nearby booming industrial town of Semarang. The zone contains pulp mills, textile works, and other big factories. Their production processes require large amounts of water, which the owners get by pumping water from the swamp on which the factories were built, she said. As the swamps empty of water, they sink. Sheltered behind breakwaters, the factories are safe enough. But the extraction of water is so intense that villages all along the shore are sinking too.

Demak officials confirmed that subsidence in some villages had reached 10 centimetres a year, thirty times the rate of sea-level rise. In 2017, Deltares, one of the Dutch firms involved in the brushwood structures, reported that 'the problems with land subsidence … are much more severe than we previously thought'.[352] Subsidence appeared to affect 'the entire 20-kilometre coastal stretch of our project.'[351,353] Alarmingly, it concluded that 'in some places, we may have reached a threshold where coastal restoration … may no longer be feasible', unless the subsidence is stopped. The brushwood structures, in other words, may not keep up.

The authorities in Semarang recognize the subsidence problem, said Khoiriyah, who had lobbied on the issue. 'They say it is not possible to forbid the pumping without finding another source of water for industry.' Eventually, the plan is to end the pumping by connecting the industries to public water-supply systems. 'But they say it will take time,' she complained. A promise has been made to call a halt before 2030. 'Meanwhile our villages are disappearing,' she said. The villagers are doing their bit, working arduously with their Dutch backers to help regenerate their mangroves and protect their coastline. But they cannot work alone. Their work can easily be undermined by others, unless an integrated approach is taken to coastal development.

What happens in Demak matters for the wider replication of the philosophy behind Building with Nature in Indonesia. The

At what price? This boy is proud of the mudskipper fish he found in his family's prawn pond. But the ponds were created at the expense of mangroves that once protected the shoreline from coastal erosion.

government is building more permeable structures along the northern Java coast, in an attempt to revive mangroves where they are most badly needed. Different types of structures are being developed for urban settings, including the country's capital Jakarta, for sandy coastlines, along vulnerable river banks, and around coral reefs. What began a decade ago in Timbulsloko in response to a local crisis is rapidly turning into a grand plan to protect a country with more coastline than any other. It would clearly be a tragedy if it were undermined by the failure of industrialists, land speculators and government bodies to allow the restoration to work. ○

CONCLUSION
GROUNDSWELL

Wetlands are among our most important global commons. As much as rainforests, they are vital to both their local inhabitants and the planet at large. Wetlands are thoroughfares and regulators of water all the way from mountain watersheds to river deltas, insuring water supplies against drought and providing space for floods when it rains. They protect coastlines, secure biodiversity, reverse land degradation, sustain fisheries, irrigate crops, neutralize pollution, moderate climate, graze livestock, nurture birdlife, and store carbon. They are refuges for wildlife and humans alike, providing food and shelter for the poor, victimized and marginalized. Their protection and restoration are essential for a habitable world.

Sometimes we can appreciate their value clearly, especially when their wildlife makes them money-earners from tourism. Even in the most remote regions, wetlands can attract large numbers of people. Visitors to the papyrus swamps of Botswana's Okavango Delta generate a tenth of the country's GDP. The Pantanal in South America has as much wildlife as the Amazon, and the Sudd swamp on the White Nile has megafauna migrations as good as any in Africa.

But such picture-perfect places are the exceptions. Most wetlands are used landscapes, mixing wildlife with often intense human exploitation. Their very importance for humanity can demote their status in debates about the global ecological restoration we need in the twenty-first century. Being neither pristine nature nor under formal ownership, their management is often neglected. The lakes of Ethiopia's Central Rift Valley are the basis of all life and human activity there; but who is in charge of them? The answer is nobody.

And where users compete – hydroelectricity versus fisheries on the rivers of South-East Asia, for instance, or irrigated agriculture versus nomadic pastoralism in the floodplains of sub-Saharan Africa – the means to make sensible trade-offs and to seek shared benefits are rarely in place. Usually the most powerful lobby wins. The iron grip of hydroelectricity generators on Lake Loktak in India, the 'mirror of Manipur', makes no sense. But it prevails.

The ecological and hydrological strengths of wetlands have come to be a weakness in the modern world. Water does not recognize fences or property boundaries, or even national borders. It flows; it permeates; it gets everywhere;

it comes and it goes. And so does its wildlife. Protecting wetlands can rarely be done by fencing them in. The good news is that wetlands can hold or release water for the benefit of those downstream. The bad news is that anything happening upstream may damage the wetland. The Pantanal, the world's largest freshwater wetland, though seemingly intact, may be close to disaster because of an agricultural boom on the grasslands of the adjacent Brazilian cerrado. Damaging the Pantanal could in turn cause huge threats from droughts, floods and much else for 2,000 kilometres downstream to the ocean.

The downstream impacts can be social and political as well as ecological. And can extend far from the wetland, sometimes even far beyond the river basin concerned. Dam operations in Mali, Niger, and Cameroon have helped trigger poverty and social breakdown in wetlands in neighbouring countries that have in turn created streams of migrants who are destabilizing governments in Western Europe.

This book is a call to action for restoring the world's wetlands, and indeed creating new ones. It is not a blueprint. Wetlands and their contexts vary too much for that. But the examples we have chosen suggest important ways forward. The prize is vital. Not just to secure the future of the world's millions of wetland species, important though that surely is, but for the improved food and water security of hundreds of millions of humans, who depend on water in wetlands or the water downstream that wetlands secure for river systems. Water security includes protection against bad floods but also the continuance of the good floods needed to sustain the wetlands themselves. It is important to be clear about the distinction.

We embrace the new ways of thinking about nature being adopted in the twenty-first century. We see wetlands as vital for human systems as well as ecosystems: as natural capital to be cherished and augmented, and as the sources of nature-based solutions to many problems conventionally dealt with by pouring concrete. A world resilient to inevitable climate change – especially to the coming hydrological turmoil from changing patterns of precipitation, evaporation and run-off, rising tides and superstorms – will be built on such solutions. Last but not least, peatlands and the blue carbon of mangroves and other coastal wetlands will be some of the best and most secure natural carbon stores in an era that may be dominated by efforts to fight climate change.

Clearly, making a success of wetlands in the twenty-first century is a wicked problem that will require policy-making based on good hydrology, good ecology, good economics, and good politics. Funding devoted only to rolling out technical solutions – whether 'building with nature' or not – will fail if the relevant stakeholders and knowledge holders are not at the table. There may be something else essential, too: ecological justice for the communities that live in the wetlands. They are stakeholders like no others. Many wetlands will only be saved and restored through being managed, controlled, and owned by the people whose livelihoods and ways of thinking have evolved over centuries to simultaneously exploit and protect their environment.

There is growing acceptance today that indigenous forest communities are the most skilled harvesters and most effective guardians of their forests. The same is true of wetlands. Several chapters in this book – on the Inner Niger Delta, the Rufiji Delta and elsewhere – address why recognizing the customary rights of wetlanders is both just and essential. The reason comes down to the wise words of Sarah Laborde of the Australian River Institute at Griffith University in Queensland, after studying the Logone Wetlands in Cameroon, 'Floodplains are complex systems with tight couplings between hydrological and social processes.'[354] The nomadic fishers and pastoralists – by sniffing the air, swapping stories over tea and taking their boats and animals far and wide to find pastures and fishing grounds – operated a smart system of wetland management and exploitation that no scientist could model and no government could operate.

When ecosystems collapse, outsiders often talk about a 'tragedy of the commons', in which resources are trashed because nobody owns them, so everybody grabs what they can while they can.[355] But often the tragedy is actually the opposite – a breakdown of old collective methods of managing those commons. Such breakdowns, moreover, are usually triggered by outside pressures to tame and own them. In the case of wetlands, that is typically by damming, dyking, and draining. We have seen that destructive force in action from China's Ruoergai Plateau to the banks of the Rhine and the Rupununi grasslands of Guyana.

So let's celebrate the 'commons'. Collective systems of management may be part of the solution, rather than part of the problem on wetlands. And let's remember that collective management can be created afresh. On the shores of Lake Prespa and elsewhere, when governments stand back, farmers, fishers, conservationists, herders, and others can reach common accords for managing wetland resources that they all rely on.

It is in this spirit that Wetlands International and its partners have sought to reach out to the inhabitants and primary users of wetlands. We have embraced the idea of 'bio-rights', offering wetland communities finance and technical expertise to help them use their knowledge to generate income while taking collective control of their environment.[350] Far from seeing wetland communities as a threat to their environments, we believe that the key to wise use of wetlands lies in empowering such communities. We see the fruits of that work in places such as Java, the Inner Niger Delta, the Mahanadi Delta and Aceh. The approach could provide a blueprint for many other wetlands and commonly owned natural resources.

Wetlands, then, are the keystones of our planet's terrestrial hydrology. Their demise, if continued, risks causing profound human suffering, social breakdown, and conflict. In an increasingly interconnected world, local breakdown can swiftly have global impacts – on migrant flows, for instance. We believe that, for such reasons, it is vital to help wetland communities to manage and control their worlds in ways that enhance their ecological and social resilience against external risks, whether from climate change or political and economic volatility.

The world needs what we might call a last ditch stand to safeguard wetlands in the Anthropocene landscapes of the twenty-first century. Wetlands are among the greatest global commons, and essential to the continued availability of water for people and nature in a crowded world. The management of these commons need not end in tragedy at the hands of their inhabitants. Far from it. For the key to their survival lies with the communities that know them best, the wetlanders themselves.

Finally, we are optimists. We see ourselves engaged not so much in a last ditch stand as a first ditch advance, in a process of revitalizing and restoring wetland ecosystems through reconnecting them with their inhabitants. Wetland restoration can never be a purely technical exercise – even if carried out by engineers and ecologists well-versed in ideas about building with nature. Ultimately, it is that reconnection between wetlands and wetlanders that is the most vital step of all. o

END NOTES

1. Wetlands International, *Downstream Voices: Wetland Solutions in Reducing Disaster Risk*, Ede, 2014, https://www.wetlands.org/publications/downstream-voices/
2. https://www.tandfonline.com/doi/abs/10.1080/02626667.2011.631014
3. Balian, E. V., Segers, H., Lévèque, C. & Martens, K. 2008. The Freshwater Animal Diversity Assessment: an overview of the results. *Hydrobiologia* 595(1): 627–37, doi:10.1007/s10750-007-9246-3
4. https://www.ramsar.org/sites/default/files/documents/library/info2007-01-e.pdf
5. https://www.ifad.org/en/web/latest/photo/asset/40799199
6. Prince, Hugh. *Wetlands of the American Midwest: A historical geography of changing attitudes*, Chicago University Press, 1997
7. http://www.e-pao.net/epSubPageExtractor.asp?src=manipur.Folks.Loktak_Folklore_Museum_aims_high_to_preserve_dying_practices_of_fishing_community
8. https://science.sciencemag.org/content/297/5583/950
9. https://www.researchgate.net/publication/284648535_Degradation_and_restoration_of_peatlands_on_the_Tibetan_Plateau
10. Interview with Gu Haijun of Sichuan Wetlands Management Center
11. Interview with the Director of Ruoergai National Nature Reserve
12. https://www.researchgate.net/publication/316166405_Restoration_of_high_altitude_peatlands_on_the_Ruoergai_Plateau
13. https://reader.elsevier.com/reader/sd/pii/S1876610211011167?token=FC96310414F232D299C95CBB-C942EEEF1
14. https://www.cambridge.org/core/books/peatland-restoration-and-ecosystem-services/restoration-of-high-altitude-peatlands-on-the-ruoergai-plateau-northeastern-tibetan-plateau-china/A638A70F028227666F25635F572067D9
15. https://www.sciencedirect.com/science/article/pii/S1876610211011167
16. https://www.researchgate.net/publication/284648535_Degradation_and_restoration_of_peatlands_on_the_Tibetan_Plateau
17. http://www.chinadaily.com.cn/china/19thcpcnationalcongress/2017-11/04/content_34115212.htm
18. https://www.chinadialogue.net/article/show/single/en/11370-Can-the-Tibetan-plateau-be-grazed-sustainably-
19. https://onlinelibrary.wiley.com/doi/abs/10.1111/jbi.12212
20. Luteyn, James L. *Páramos: A Checklist of Plant Diversity, Geographical Distribution, and Botanical Literature*, Memoirs of the New York Botanical Garden Volume 84, 1999
21. Molina López, Jorge et al. *Forging an Alliance Negotiation of Interests and Conflict Transformation by Civil Society, Public and Private Stakeholders for the Protection of the Santurbàn-Sisavita Highland*, GIZ, 2013
22. https://www.rvo.nl/sites/default/files/2018/11/Water%20Management%20-%20G2%20-%20Water%20Management.pdf
23. https://www.prensarural.org/spip/spip.php?article1280
24. https://www.researchgate.net/publication/308777838_The_Paramo_Ecosystem_of_Costa_Rica's_Highlands
25. http://unfccc.int/resource/docs/natc/colnc1.pdf
26. Herzog, S.K. et al. (eds). *Climate Change and Biodiversity in the Tropical Andes*. 2011, SCOPE/IAI, chap.12, http://www.infobosques.com/descargas/biblioteca/214.pdf
27. https://www.thegef.org/project/conservation-biodiversity-paramo-northern-and-central-andes
28. https://www.jstor.org/stable/1733739?seq=1#page_scan_tab_contents
29. http://www.fao.org/3/I9159EN/i9159en.pdf
30. https://www.researchgate.net/publication/247328728_Raised_Fields_and_Sustainable_Agriculture_in_the_Lake_Titicaca_Basin_of_Peru
31. https://www.conservationmagazine.org/2008/07/virginity-lost/
32. https://www.tandfonline.com/doi/abs/10.1080/11263504.2013.870250?journalCode=tplb20
33. http://datazone.birdlife.org/site/factsheet/loktak-lake-and-keibul-lamjao-national-park-iba-india/text
34. http://www.iosrjournals.org/iosr-jhss/papers/Vol.%2023%20Issue5/Version-6/F2305066772.pdf
35. https://lib.ohchr.org/HRBodies/UPR/Documents/Session1/IN/COHR_IND_UPR_S1_2008anx_Annex%20XXI_Operation%20loktak.pdf
36. http://www.omct.org/statements/india/2018/09/d25050/
37. Kumar, Ritesh & Meitei, Ng. Sanajaoba. 2011. Water Management Plan for Loktak: Balancing Human and Ecological Needs. *Loktak Newsletter* 6:14–20 (Ritesh Kumar of Wetlands International; Ng. Sanajaoba Meitei of Loktak Development Authority.)
38. http://edepot.wur.nl/19397
39. https://sherethiopie.com/en/about-us/
40. https://www.sciencedirect.com/science/article/pii/S2214581817300988
41. https://www.wetlands.org/blog/can-this-be-the-next-nakuru/
42. https://www.omicsonline.org/open-access/greening-a-tropical-abijatashala-lakes-national-park-ethiopia--a-review-2157-7625-1000179.php?aid=70293
43. https://www.researchgate.net/publication/299490337_Greening_a_Tropical_Abijata-Shala_Lakes_National_Park_Ethiopia_-_A_Review
44. http://images.agri-profocus.nl/upload/Experience_from_Meki_Batu1460654164.pdf
45. https://www.ft.com/content/7be57c9a-6f01-11e8-92d3-6c13e5c92914
46. https://www.hrw.org/news/2017/02/14/ethiopia-dams-plantations-threat-kenyans
47. http://www.whpress.co.uk/EH/papers/1536.pdf
48. http://archive.wetlands.org/Portals/0/publications/Report/QuickScan%20of%20peatlands%20in%20central%20and%20eastern%20europe%20CROP.pdf
49. http://mires-and-peat.net/media/map19/map_19_14.pdf
50. https://www.theguardian.com/world/2010/aug/06/russia-fires-moscow
51. https://www.abc.net.au/news/2010-09-18/russian-heatwave-killed-11000-people/2265184
52. https://www.wetlands.org/publications/restoring-peatlands-in-russia/

53. https://unfccc.int/climate-action/momentum-for-change/planetary-health/restoring-peatlands-in-russia-i-russia
54. https://www.kfw.de/stories/environment/climate-change/moore-russland/
55. https://agupubs.onlinelibrary.wiley.com/doi/10.1029/2004GL022025
56. Pearce, Fred. 1993. The Scandal of Siberia, *New Scientist*. London, (27 November 1993), 28–33
57. https://www.newscientist.com/article/mg18725124-500-climate-warning-as-siberia-melts/
58. https://www.nature.com/articles/srep17951
59. http://science.sciencemag.org/content/303/5656/353
60. https://agupubs.onlinelibrary.wiley.com/doi/full/10.1029/2008GB003327
61. https://agupubs.onlinelibrary.wiley.com/doi/full/10.1029/2011EO090001
62. https://earthdata.nasa.gov/learn/sensing-our-planet/leaking-lakes
63. https://www.nature.com/articles/d41586-019-01313-4
64. https://www.newscientist.com/article/2211013-huge-arctic-fires-have-now-emitted-a-record-breaking-amount-of-co2/
65. https://www.nature.com/articles/s41561-018-0174-9
66. http://www.iucn-uk-peatlandprogramme.org/sites/www.iucn-uk-peatlandprogramme.org/files/IUCN%20UK%20Commission%20of%20Inquiry%20on%20Peatlands%20Full%20Report%20spv%20web_1.pdf
67. https://www.ingentaconnect.com/content/whp/eh/pre-prints/content-whp_eh_1536
68. https://www.iucn-uk-peatlandprogramme.org/projects/flows-bringing-life-back-bogs
69. http://www.iucn-uk-peatlandprogramme.org/news-and-events/news/humble-peatlands-save-uk-%C2%A3millions
70. https://www.wetlands.org/publications/briefing-paper-accelerating-action-to-save-peat-for-less-heat/
71. Hooijer, A., Silvius, M., Wösten, H. and Page, S. 2006. *PEAT-CO2, Assessment of CO2 emissions from drained peatlands in SE Asia*. Delft Hydraulics report Q3943
72. https://www.iucn.org/content/moorfutures-%E2%80%93-how-regional-carbon-credits-peatland-rewetting-can-help-nature-conservation
73. http://science.sciencemag.org/content/362/6420/1222
74. https://www.theguardian.com/world/2018/nov/27/ireland-closes-peat-bogs-climate-change
75. https://www.independent.ie/business/farming/forestry-enviro/environment/bord-na-mna-digging-deep-to-rewet-thousands-of-acres-of-bog-36998517.html
76. https://www.responsiblyproducedpeat.org/en/what-we-do
77. https://link.springer.com/article/10.1007/s10745-018-9977-y
78. Darby, H.C. *The Changing Fenland*, Cambridge University Press, 1983
79. Purseglove, Jeremy. *Taming the Flood*, William Collins, 1988
80. https://www.tandfonline.com/doi/abs/10.1179/jrl.2006.2.2.109?journalCode=yjrl19
81. https://search.proquest.com/openview/535d-7b3734eb6612/1?pq-origsite=gscholar&cbl=1460
82. https://threeacresandacow.co.uk/2014/07/the-fowlers-complaint-the-powtes-complaint-1611-the-fens-trad/
83. Goulden, Glenda. *Foul Deeds & Suspicious Deaths In & Around the Fens*, Wharncliffe True Crime, 2008
84. https://www.newscientist.com/article/mg17723865-800-ghosts-of-the-great-eel-war/
85. http://www.greatfen.org.uk/heritage/engineering
86. http://www.greatfen.org.uk/about/future
87. Dickens, Charles. *Our Mutual Friend*, Chapman & Hall, 1864
88. McNeill, John. *Something New Under the Sun*, Allen Lane, 2000
89. https://www.perc.org/2002/03/01/who-drained-the-everglades/
90. Shapiro, Judith. *Mao's War Against Nature*, Cambridge University Press, 2001
91. https://www.chinatravel.com/kunming-attraction/haigeng-park/
92. https://www.academia.edu/4173042/A_conservation_agenda_for_the_Pantanal_s_biodiversity
93. https://cdr.lib.unc.edu/indexablecontent/uuid:2476f22b-9237-4bc4-8304-d62f39a3b6bd
94. http://www.scielo.br/scielo.php?script=sci_arttext&pid=S1519-69842011000200012
95. http://discovery.ucl.ac.uk/1551634/1/RChiaravalloti_PhD_16_04.pdf
96. http://ecoa.org.br/area-de-protecao-e-ecoturismo-no-pantanal-inclui-um-dos-sitios-ramsar/
97. http://macaulay.webarchive.hutton.ac.uk/pantanal/
98. http://www.pousalegre.com.br/
99. https://www.sei.org/about-sei/press-room/press-releases/soy-trade-from-brazils-cerrado-driving-climate-emissions/
100. https://pdfs.semanticscholar.org/2218/b72369312745e5c1924d-64d5d67f86425609.pdf
101. http://www.observatoriodoagronegocio.com.br/page41/files/que-mpagaconta.pdf
102. http://www.wetlands.org/news/new-ten-year-corredor-azul-programme-south-america/
103. http://www.forestpeoples.org/en/topics/environmental-governance/publication/2015/where-they-stand
104. http://d2ouvy59p0dg6k.cloudfront.net/downloads/wwf_bsrs_booklet_low_res_2.pdf
105. https://www.forestpeoples.org/en/environmental-governance/press-release/2017/press-release-wapichan-people-expose-rights-violations
106. http://citeseerx.ist.psu.edu/viewdoc/download?doi=10.1.1.458.7313&rep=rep1&type=pdf
107. Poole, Colin. *Tonle Sap: The Heart of Cambodia's Natural Heritage*, River Books, 2005
108. http://www.jswconline.org/content/73/3/60A
109. http://www.fao.org/fi/oldsite/FCP/en/KHM/profile.htm
110. http://www.mrcmekong.org/assets/Publications/report-management-develop/Mek-Dev-No2-Mek-River-Biodiversifiisheries-in.pdf
111. https://chinadialogue.net/article/show/single/en/10901-Can-Chinese-reciprocity-protect-the-Mekong-
112. https://www.sei.org/publications/sediment-mekong-river/
113. https://www.researchgate.net/publication/228450282_Tonle_Sap_pulsing_system_and_fisheries_productivity
114. http://www.rmz-mg.com/letniki/rmz52/rmz52_0087-0089.pdf
115. http://www.pnas.org/content/115/47/11891
116. https://iopscience.iop.org/article/10.1088/1748-9326/10/1/015001/meta
117. https://www.wetlands.org/blog/announcement-build-guinea-dam-bypasses-regional-collaborative-process/
118. http://news.bbc.co.uk/1/hi/world/africa/7959444.stm
119. http://www.mursi.org/pdf/Meles%20Jinka%20speech.pdf
120. https://papers.ssrn.com/sol3/papers.cfm?abstract_id=2406852
121. https://www.nature.com/articles/ncomms15697
122. Pearce, Fred. 1992. Death of an Oasis, *Audubon*, (May/June 1992)
123. http://yahyadanmato.blogspot.com/2017/02/need-for-restoration-of-bade-annual.html
124. https://www.tandfonline.com/doi/abs/10.1080/02626667.2011.629787
125. http://www.un.org/africarenewal/magazine/april-2012/africa%E2%80%99s-vanishing-lake-chad
126. https://afrosai-e.org.za/uploads/afrosai_intohost_co_za/cms/files/environmental_audit_on_the_drying_up_of_lake_chad_nigeria.pdf
127. https://www.giz.de/de/downloads/giz2015-en-joint-environmental-audit-report-lake-chad.pdf
128. https://www.tandfonline.com/doi/abs/10.1080/01431161.2013.827813
129. http://wedocs.unep.org/bitstream/handle/20.500.11822/8398/-Lake%20Chad%20Basin-GIWA%20Regional%20Assessment%2043-20043745.pdf?sequence=3&isAllowed=y
130. https://openaccess.leidenuniv.nl/handle/1887/4290

131 https://books.google.co.uk/1cQSi01gC&pg=PA5&lpg=PA5&d-q=%E2%80%9Cdiminished+rather+than+improved+the+living+standards+and+economy+of+the+region+as+a+whole%E2%80%9D&source=bl&ots=Uv7pyit-EDY&sig=vpkkXIefp4nicENo142UzJz6G5c&hl=en&sa=X-&ved=2ahUKEwjbytGt_MzfAhUIUhoKHc39A8AQ6AEwA-noECAMQAQ#v=onepage&q=%E2%80%9Cdiminished%20rather%20than%20improved%20the%20living%20standards%20and%20economy%20of%20the%20region%20as%20a%20whole%E2%80%9D&f=false

132 Loth, Paul. *The Return of the Water: Restoring the Waza Logone Floodplain in Cameroon*, IUCN, 2004, https://portals.iucn.org/library/sites/library/files/documents/wtl-030.pdf

133 https://www.crisisgroup.org/africa/west-africa/nigeria/252-herders-against-farmers-nigerias-expanding-deadly-conflict

134 https://www.iom.int/news/over-26-million-displaced-lake-chad-basin-iom

135 http://www.dw.com/en/lake-chad-recedes-to-catastrophic-levels/a-18879406

136 http://www.dailymail.co.uk/wires/afp/article-3308904/Africas-Lake-Chad-fuel-new-migrant-crisis-UN.html

137 http://allafrica.com/view/group/main/main/id/00048990.html

138 https://www.bbc.co.uk/news/world-africa-43500314

139 https://onlinelibrary.wiley.com/doi/abs/10.1111/j.1747-6593.1996.tb00076.x

140 https://www.wetlands.org/publications/water-shocks-wetlands-human-migration-sahel/

141 https://iwlearn.net/resolveuid/655fb506257eb4736597193d-994f257e

142 https://www.wetlands.org/publications/managing-malis-wetland-wealth-for-people-and-nature/

143 http://www.iwmi.cgiar.org/Publications/Books/PDF/wetlands-and-people.pdf

144 http://www.altwym.nl/uploads/pdf/133Executive%20summary%20-%20The%20Niger,%20a%20lifeline.pdf

145 https://www.researchgate.net/publication/293796786_Bio-rights_in_theory_and_practice_A_financing_mechanism_for_linking_poverty_alleviation_and_environmental_conservation

146 Zwarts, L., Cisse, N. & Diallo, M. 2005a. Hydrology of the Upper Niger. In: L. Zwarts, P. van Beukering, B. Kone & E. Wymenga, eds. *The Niger, a lifeline: Effective water management in the Upper Niger Basin*. RIZA

147 https://theconversation.com/malis-volatile-mix-of-communal-rivalries-and-a-weak-state-is-fuelling-jihadism-114442

148 https://www.wsj.com/articles/african-dream-of-europe-turns-into-a-nightmare-1451405269

149 https://www.wetlands.org/blog/announcement-build-guinea-dam-bypasses-regional-collaborative-process/

150 https://www.wetlands.org/news/new-irrigation-plans-threaten-food-production-inner-niger-delta/

151 Baker, Samuel White. *Ismailia: A Narrative of the Expedition to Central Africa for the Suppression of the Slave Trade Organized by Ismail Khedive of Egypt*, Library of Alexandria, 1879

152 https://www.researchgate.net/publication/258644225_Flood_Pulsing_in_the_Sudd_Wetland_Analysis_of_Seasonal_Variations_in_Inundation_and_Evaporation_in_South_Sudan

153 http://blogs.reuters.com/world-wrap/2010/11/22/driving-sudan-towards-paradise/

154 https://news.nationalgeographic.com/2017/02/sudd-south-sudan/

155 Thesiger, Wilfred. *The Marsh Arabs*, Longmans, Green, 1964

156 Haigh, Frank. *Control of the Rivers of Iraq*. Iraq Irrigation Development Commission, 1951 (accessed in Institution of Civil Engineers Library, London)

157 https://www.newscientist.com/article/mg13818691-800-draining-life-from-iraqs-marshes-saddam-hussein-is-using-an-old-idea-to-force-the-marsh-arabs-from-their-home/

158 https://na.unep.net/siouxfalls/publications/meso.pdf

159 http://www.unesco.org/new/en/member-states/single-view/news/the_marshlands_of_iraq_inscribed_on_unescos_world_heritag/

160 https://www.newscientist.com/article/mg21829130-200-we-can-save-iraqs-garden-of-eden/

161 https://www.researchgate.net/publication/291002277_The_rectification_of_the_river_Rhine_by_Johann_Gottfried_Tulla

162 Cioc, Mark. *The Rhine: An Eco-biography 1815–2000*. University of Washington Press, 2002

163 Bernhardt, Christoph. *PCCP Processes in History: The model of the Upper Rhine Region*. UNESCO, 2001

164 https://www.welt.de/vermischtes/article169697021/Dieser-Mann-hatte-einen-irren-Plan-fuer-den-Rhein.html

165 Hollis, G.E. & Jones, T.A. 1991. Europe and the Mediterranean Basin: in M. Finlayson & M. Moser, (eds), *Wetlands*. International Wildfowl & Wetlands Research Bureau, 27–56

166 https://www.researchgate.net/publication/320689041_Flussauen_in_Deutschland_Erfassung_und_Bewertung_des_Auenzustandes

167 http://publicaties.minienm.nl/documenten/guidelines-for-rehabilitation-and-management-of-floodplains-ecol

168 Murray, Donald S. *The Dark Stuff: Stories from the Peatlands*, Bloomsbury, 2018

169 https://www.ruimtevoorderivier.nl/depoldering-noordwaard/

170 https://www.hydrol-earth-syst-sci.net/7/358/2003/

171 https://www.bmu.de/en/pressrelease/trittin-presents-draft-flood-control-act/

172 https://e360.yale.edu/features/a_successful_push_to_restore_europes_long-abused_rivers

173 http://www.greetingsfromsaltonsea.com/flood.html

174 https://www.kpbs.org/news/2018/mar/21/how-dying-lake-california-factors-colorado-rivers-/

175 https://eu.redding.com/story/life/2017/03/23/shrinking-salton-sea-threatens-wildlife/99412070/

176 http://www.sci.sdsu.edu/salton/NiilerSaltonSeaResort.html

177 Pearce, Fred. *When The Rivers Run Dry*, Granta, 2019

178 http://pacinst.org/app/uploads/2014/09/PacInst_HazardsToll.pdf

179 https://www.nature.com/articles/nature20584

180 https://www.theguardian.com/cities/2016/mar/09/kolkata-wetlands-india-miracle-environmentalist-flood-defence

181 Pearce, Fred & Crivelli, Alain. *Characteristics of Mediterranean Wetlands*, Tour du Valat, 1994.

182 http://www.andalucia.com/environment/protect/alto-guadalquivir.htm

183 https://www.wwf.de/fileadmin/fm-wwf/Publikationen-PDF/Environmental_flows_in_the_marsh_of_the_National_Park_of_Donana.pdf

184 https://www.scidev.net/global/farming/multimedia/the-rich-diversity-of-birds-in-rice-field-ecosystems.html

185 http://www.bioone.org/doi/abs/10.1675/063.033.s101

186 https://www.japantimes.co.jp/life/2011/03/13/environment/traditional-paddies-are-great-ecosystems/#.XEeJOlyeSUk

187 https://freshwaterhabitats.org.uk/habitats/pond/

188 http://www.countrysidesurvey.org.uk/sites/default/files/CS_UK_2007_TR7%20-%20Ponds%20Report.pdf

189 https://theconversation.com/ponds-can-absorb-more-carbon-than-woodland-heres-how-they-can-fight-climate-change-in-your-garden-110652

190 Pearce, Fred. *The New Wild*, Icon Books, London 2015.

191 https://www.citylab.com/solutions/2016/08/the-dutch-are-cleaning-up-their-lakes-with-artificial-islands/494000/

192 Catsadorakis, Giogios. *Prespa: A Story for Man and Nature*, Society for the Protection of Prespa, 1999

193 https://www.theguardian.com/world/2018/jun/17/macedonia-greece-dispute-name-accord-prespa

194 http://www.spp.gr/fish_biodiversity/EN/eBook.data/02_the_fish_of_prespa.html

195 https://www.newscientist.com/article/dn28693-europes-oldest-lake-faces-destruction-to-make-way-for-tourists/

196 https://www.wetland-ecology.nl/sites/cwe.nioo.knaw.nl/files/Apostolova%20et%20al.%202016.pdf

197 http://sws.org/images/News/LakeOhrid2015.pdf

198 http://www.sws.org/images/chapters/europe/Declaration.pdf

199 https://ohridsos.files.wordpress.com/2018/04/studenchishte-proposal_ohridsos-lq.pdf

200 http://www.ecoalbania.org/new-hydropower-report-dam-tsunami-in-the-balkans-is-speeding-up/

201 https://riverwatch.eu/en/balkanrivers/news/victory-brave-women-kru%C5%A1%C4%8Dica

202 https://www.theguardian.com/world/2014/oct/01/satellite-images-show-aral-sea-basin-completely-dried
203 https://indepthnews.net/index.php/archive-search/central-asia/2289-aral-sea-promises-to-rise-like-phoenix-from-the-ashes
204 http://www.caee.utexas.edu/prof/mckinney/ce385d/papers/atanizaova_wwf3.pdf
205 https://agupubs.onlinelibrary.wiley.com/doi/10.1002/2014GL060641
206 https://earthobservatory.nasa.gov/features/hamoun
207 https://www.sciencemag.org/news/2018/02/can-iran-and-afghanistan-cooperate-bring-oasis-back-dead
208 https://www.timesofisrael.com/on-land-crumbled-by-sinkholes-dead-sea-locals-try-to-shore-up-their-livelihoods/
209 https://news.un.org/en/story/2001/09/15462-6-million-un-backed-project-pumps-life-back-jordanian-oasis
210 https://www.engr.washington.edu/node/1799
211 https://onlinelibrary.wiley.com/doi/abs/10.1111/mms.12181
212 https://www.researchgate.net/publication/297890380_When_the_lakes_run_dry
213 https://www.sciencedirect.com/science/article/pii/S0380133016301307
214 http://www.ir.undp.org/content/iran/en/home/presscenter/articles/2017/03/22/lake-urmia-comes-back-to-life-slowly-but-surely.html
215 O'Hanlon, Redmond. Congo Journey, Penguin, 1997
216 https://www.sciencedaily.com/releases/2017/01/170111132812.htm
217 https://nerc.ukri.org/planetearth/stories/1617/
218 http://eprints.whiterose.ac.uk/92620/1/10.1088.1748-9326.9.12.124017.pdf
219 www.biogeosciences.net/7/1505/2010
220 https://onlinelibrary.wiley.com/doi/abs/10.1111/gcb.13689
221 https://www2.le.ac.uk/departments/geography/research/projects/tropical-peatland/downloadable-resources/peatland-drainage
222 https://www.researchgate.net/publication/317386047_Managing_peatlands_in_Indonesia_Challenges_and_opportunities_for_local_and_global_communities
223 http://www.cifor.org/publications/pdf_files/infobrief/6449-infobrief.pdf
224 https://www.pnas.org/content/early/2017/10/11/1710465114
225 Hooijer,A., Page, S. Jauhiainen, J. LeeW.A., Lu, X.X. & Anshari, G. 2012. Subsidence and carbon loss in drained tropical peatlands. Biogeosciences 9, 1053–1071
226 Wijedasa, L.S. et al. 2016. Denial of long-term issues with agriculture on tropical peatlands will have devastating consequences. In: Letters to Editor, Global Change Biology, (Sept. 2016)
227 https://news.mongabay.com/2018/05/indonesian-government-wants-to-turn-haze-causing-mega-rice-project-around/
228 https://grist.org/business-technology/from-cutting-down-rainforests-to-restoring-them-a-company-changes-course/
229 https://theconversation.com/rising-seas-to-keep-humans-safe-let-nature-shape-the-coast-107837
230 https://en.wikipedia.org/wiki/Battle_of_Rufiji_Delta
231 https://core.ac.uk/download/pdf/4834267.pdf
232 http://www.academia.edu/10647256/The_REDD_Menace_Resurgent_Protectionism_in_Tanzanias_Mangrove_Forests
233 https://ejatlas.org/conflict/protest-against-commercial-shrimp-farming-in-rujifi-delhi-tanzania
234 https://www.equatorinitiative.org/wp-content/uploads/2017/05/case_1348164376.pdf
235 http://www.xinhuanet.com/english/2018-09/19/c_137477274.htm
236 https://redd-monitor.org/2012/05/09/wwf-scandal-part-3-corruption-and-evictions-in-tanzania/
237 http://coastalforests.tfcg.org/pubs/REMP%2007%20TR%206%20vol%202%20Selection%20of%20Four%20Additional%20Pilot%20Villages.pdf
238 https://www.sciencedirect.com/science/article/abs/pii/S0959378011001932
239 https://www.maf-uk.org/story/maf-brings-help-to-the-isolated-rufiji-delta
240 https://www.tandfonline.com/doi/abs/10.1080/02626667.2013.827792
241 https://www.cambridge.org/core/journals/environmental-conservation/article/environmental-impacts-of-the-proposed-stieglers-gorge-hydropower-project-tanzania/5C06F6165883FAA72BC6538B8D35DEE2
242 https://www.worldwildlife.org/publications/the-true-cost-of-power-the-facts-and-risks-of-building-stiegler-s-gorge-hydropower-dam-in-selous-game-reserve-tanzania
243 Environmental impact assessment of the Stiegler's Gorge hydropower project, Tanzania, University of Dar es Salaam, 2018
244 https://www.ncbi.nlm.nih.gov/pubmed/17562084
245 http://www.globalconstructionreview.com/news/tanzania-build-controversial-stieglers-gorge-dam-c/
246 https://www.egyptindependent.com/tanzanian-president-egypts-electricity-minister-lay-corner-stone-for-stieglers-gorge-dam/
247 McPhee, John. The Control of Nature, Farrar, Straus and Giroux, 1989
248 Twain, Mark. Life on the Mississippi, Harper & brothers, 1901
249 https://www.organicconsumers.org/news/la-plan-reclaim-land-would-divert-mississippi
250 https://eu.usatoday.com/story/news/nation/2013/02/23/leeville-levees-louisiana/1939695/
251 http://www.isledejeancharles.com/
252 http://news.bbc.co.uk/1/hi/world/americas/4393852.stm
253 http://coastal.la.gov/wp-content/uploads/2016/08/2017-MP-Book_Single_Combined_01.05.2017.pdf
254 https://www.pakistantoday.com.pk/2011/07/18/keti-bunder-%E2%80%93-the-town-that-never-became-the-third-port-of-sindh/
255 https://www.reuters.com/article/us-pakistan-climatechange-migration/pakistans-coastal-villagers-retreat-as-seas-gobble-land-idUSKBN0KI0XI20150109?feedType=RSS
256 https://jepsl.sljol.info/articles/abstract/10.4038/jepsl.v3i1.7310/
257 http://www.lead.org.pk/hr/attachments/Compandium/04_Environmental_Rights/Death_of_the_Indus_Delta.pdf
258 https://www.theguardian.com/world/2010/oct/05/pakistan-flood-waters-indus-delta
259 https://www.sciencedirect.com/science/article/pii/S0921818103000237
260 https://www.sciencedirect.com/science/article/pii/S1687428516000108
261 Leopold, Aldo. A Sand County Almanac & Other Writings on Conservation and Ecology, Library America, 2013
262 https://www.issuelab.org/resource/a-delta-once-more-restoring-riparian-and-wetland-habitat-in-the-colorado-river-delta.html
263 https://raisetheriver.org/
264 https://www.nfwf.org/delta/Pages/home.aspx
265 https://www.nrdc.org/onearth/colorado-river-delta-proof-natures-resiliency
266 https://www.newscientist.com/article/mg20927942-600-intensive-logging-created-new-englands-rich-wetlands/
267 https://www.researchgate.net/publication/236977168_Man_made_deltas
268 https://onlinelibrary.wiley.com/doi/abs/10.1002/%28SICI%291099-1646%28199601%2912%3A1%3C51%3A%3AAID-RRR376%3E3.0.CO%3B2-I
269 Interview with Carles Ibanez of the Institute of Agri-food Research and Technology, Catalonia
270 https://www.theguardian.com/world/2018/aug/17/kerala-floods-death-toll-rescue-effort-india
271 Roy, Arundhati. The God of Small Things, IndiaInk, 1997
272 https://www.thehindu.com/news/national/kerala/rs17698-crore-sanctioned-for-kole-land-development/article4955591.ece
273 https://uk.reuters.com/article/uk-india-floods-insight/did-keralas-dams-exacerbate-indias-once-in-century-floods-idUKKCN1MK2XE
274 https://www.deccanchronicle.com/nation/current-affairs/120818/shutters-should-have-been-lifted-early-prof-ej-james.html
275 http://www.irrigation.kerala.gov.in/images/latest/flood_2018_case_study.pdf

276 https://www.wetlands.org/blog/kerala-floods-from-the-eyes-of-wetlands/
277 https://www.theguardian.com/world/2018/oct/29/venice-experiences-worst-flooding-since-2008
278 https://whc.unesco.org/en/list/394
279 Da Mosto, Jane & Fletcher, Caroline. The Science of Saving Venice, Umberto Allemandi, 2004
280 https://scroll.in/article/902011/why-venices-plan-to-tackle-worsening-floods-is-being-criticised
281 https://www.nature.com/articles/d41586-018-07372-3
282 https://www.jstor.org/stable/25177065?seq=1#page_scan_tab_contents
283 https://www.nytimes.com/2012/11/04/nyregion/protecting-new-york-city-before-next-time.html
284 http://www.anthropocenemagazine.org/2014/10/the-future-will-not-be-dry/
285 https://urbandemos.nyu.edu/2019/03/11/les-coastal-resiliency-plan/
286 https://esajournals.onlinelibrary.wiley.com/doi/pdf/10.1002/fee.1494
287 http://scc.ca.gov/webmaster/ftp/pdf/sccb-b/2007/0711/0711Board11b_Greening_the_Bay_Ex1.pdf
288 https://www.greenbiz.com/article/chinas-sponge-cities-aim-reuse-most-rainwater
289 https://www.wetlands.org/news/netherlands-panama-governments-seek-help-collaboration-water/
290 https://www.bbc.co.uk/news/world-latin-america-31351625
291 http://news.bbc.co.uk/1/hi/business/4528332.stm
292 http://science.sciencemag.org/content/310/5748/643
293 https://www.researchgate.net/publication/238077579_Greenbelt_Tsunami_Prevention_in_South-Pacific_Region
294 https://www.conservationgateway.org/ConservationPractices/Marine/crr/library/Pages/Mangroves-coastal-defence.aspx
295 https://www.pnas.org/content/108/46/18612
296 https://www.wetlands.org/publications/mangrove-restoration-to-plant-or-not-to-plant/
297 https://link.springer.com/article/10.1007/s10584-010-0003-7
298 https://www.weadapt.org/knowledge-base/disaster-resilience/mangroves-for-coastal-defence
299 http://iopscience.iop.org/article/10.1088/1748-9326/aabe1c/meta
300 https://journals.plos.org/plosone/article?id=10.1371/journal.pone.0010095
301 https://www.newscientist.com/article/dn27498-sri-lanka-first-nation-to-promise-full-protection-of-mangroves/
302 https://onlinelibrary.wiley.com/doi/abs/10.1111/rec.12492
303 https://link.springer.com/article/10.1007/s12237-015-9993-8
304 https://www.wetlands.org/blog/restoring-mangroves-abandoned-rice-fields-guinea-bissau/
305 https://www.wetlands.org/publications/conserving-biodiversity-cacheu-mangroves-national-park-guinea-bissau/
306 https://www.wetlands.org/blog/living-with-floods-in-the-mahanadi-delta-india/
307 http://www.pnas.org/content/106/18/7357
308 https://www.voanews.com/a/worst-of-indian-cyclone-is-over/1768632.html
309 https://indianexpress.com/article/news-archive/web/month-after-cyclone-phailin-odisha-looking-at-devastation-on-the-scale-of-1999/
310 Partners for Resilience Narrative Progress Report – India, Red Cross Climate Center, 2012
311 https://english.rvo.nl/sites/default/files/2018/04/Call%20for%20Action%20-%20final.pdf
312 https://www.royalhaskoningdhv.com/en-gb/news-room/news/alleviate-water-risks-and-enhance-resilience-in-khulna-bangladesh/8318
313 https://earthobservatory.nasa.gov/world-of-change/PadmaRiver
314 Interview with John Pethick of Queen's University Belfast
315 Hydro-Meteorological Trends in Southwest Coastal Bangladesh: Perspectives of Climate Change and Human Interventions, American Journal of Climate Change, doi:10.4236/ajcc.2013.21007
316 &, J.D.Rapid Rise in Effective Sea-level in Southwest Bangladesh: Its causes and contemporary rates, Global and Planetary Change, doi:10.1016/j.gloplacha.2013.09.019
317 https://link.springer.com/article/10.1007/s11027-015-9640-5
318 https://www.waddenzee.nl/fileadmin/content/Dossiers/Natuur_en_Landschap/pdf/Brochure10yrKwelderherstel_NFB_Eng_okt2014.pdf
319 https://www.waddenzee.nl/fileadmin/content/Dossiers/Natuur_en_Landschap/pdf/From_polder_to_saltmarsh_ENG.pdf
320 https://journals.plos.org/plosone/article?id=10.1371/journal.pone.0027374
321 https://link.springer.com/article/10.1007/s10021-018-0332-3
322 https://hansard.parliament.uk/Commons/2004-03-12/debates/dc1b07f9-b057-461a-bee1-64a795ce7391/ThamesEstuary
323 https://www.researchgate.net/publication/225767379_25_years_of_salt_marsh_erosion_in_Essex_Implications_for_coastal_defence_and_nature_conservation
324 https://www.rspb.org.uk/our-work/our-positions-and-casework/casework/cases/wallasea-island/
325 http://www.bioone.org/doi/pdf/10.1579/0044-7447%282008%2937%5B241%3ATVOCWF%5D2.0.CO%3B2
326 https://delawareestuary.s3.amazonaws.com/pdf/Living%20Shorelines/living_shorelines_best_practices.pdf and
327 https://e360.yale.edu/features/why_restoring_wetlands_is_more_critical_than_ever
328 https://www.sciencemag.org/news/2015/10/china-s-vanishing-coastal-wetlands-are-nearing-critical-red-line
329 https://www.wetlands.org/news/arcadia-fund-helps-us-support-restoration-management-critical-habitats-migratory-waterbirds-yellow-sea/
330 https://www.koreaexpose.com/saemangeum-wetland-destroyed-korea/
331 https://www.pv-magazine.com/2018/11/01/korea-to-build-worlds-largest-solar-park/
332 http://mires-and-peat.net/media/map16/map_16_01.pdf
333 https://www.nytimes.com/2000/10/20/us/everglades-restoration-plan-passes-house-with-final-approval-seen.html?scp=1&st=nyt
334 http://dels.nas.edu/Report/Progress-Toward-Restoring-Everglades/23672?bname=banr
335 https://www.sciencedirect.com/science/article/pii/S1462901116306098
336 http://www.grida.no/resources/12529
337 http://thebluecarboninitiative.org/blue-carbon/
338 https://www.ncbi.nlm.nih.gov/pubmed/28544444
339 http://www.cifor.org/publications/pdf_files/Books/BMurdiyarso1401.pdf
340 https://www.nap.edu/read/25259/chapter/4
341 https://www.nature.com/articles/d41586-018-00018-4
342 http://thebluecarboninitiative.org/
343 https://www.cifor.org/library/7058/
344 https://royalsocietypublishing.org/doi/10.1098/rsbl.2018.0773
345 https://www.nap.edu/read/25259/
346 https://asiapacificreport.nz/2017/11/18/tiny-timbulsloko-fights-back-in-face-of-indonesias-ecological-disaster/
347 http://iopscience.iop.org/article/10.1088/1755-1315/139/1/012035
348 https://news.mongabay.com/2017/07/photos-where-once-were-mangroves-javan-villages-struggle-to-beat-back-the-sea/
349 https://magazine.boskalis.com/issue04/eco-shaping-the-future
350 https://buildingwithnatureindonesia.exposure.co/building-with-nature-indonesia-3
351 https://www.researchgate.net/publication/293796786_Biorights_in_theory_and_practice_A_financing_mechanism_for_linking_poverty_alleviation_and_environmental_conservation
352 https://buildingwithnatureindonesia.exposure.co/building-with-nature-indonesia-1
353 https://www.wetlands.org/news/first-design-workshop-semarang-develop-solutions-citys-sinking-flooding-challenges/
354 https://www.researchgate.net/publication/323400243_The_interplay_of_top-down_planning_and_adaptive_self-organization_in_an_African_floodplain
355 http://science.sciencemag.org/content/162/3859/1243

INDEX

Page references in *italics* indicate images.

A

Abera, Yadecha 57
Abijata, Lake, Ethiopia *50*, 51, *54–5*, 56–7
accidental wetlands 13, 160, 162–72, *162*, *165*, *166–7*, *170*, *171*, 238
Aceh, Indonesia 199, 239–48, *240–1*, *242*, *243*, *244–5*, *246–7*, 283, 291
Advincola, Joison 100–1
Ahmed, Abiy 57
Albrecht, Christian 181
Alier, Abel 142
All-American Canal, U.S. 164
Alongi, Daniel 246
Alwash, Azzam 148
Amazon rainforest, South America 33, 38, 96, 109, 147
Amazon River, South America 33, 80, 112, 190, 288
Amu Darya River, Asia 183, 184
Andes mountains, South America *14–15*, 16, 21, 31–8, *34–5*, *37*
Angkor Wat, Cambodia 10, 118–19, *120–1*
Ansar, Atif 124
Anthony, Katey Walter 74
Apostolova, Nadezda 182
Aquidauana, River, Brazil 105
Aral Sea, Asia 6, 14, 147, *158–9*, 160, *161*, 183–8, *184*, *185*, *186–7*
Army Corps of Engineers, U.S. 89, 213, 268
artificial wetlands 162–72, *162*, *165*, *167–8*, *170*, *171*, 238
Asia Pulp & Paper 194
Astra, Apri Susanto 285
Avery, Sean 60
Azraq oasis, Jordan 188
Aztec Empire 36–8, *37*

B

Baker, Sir Samuel White 141
Bakker, Chris 259, 260, 261, 264
Barbier, Edward 127
Barn Elms Wetland Centre 238
Bay of Bengal, Asia 11, 198, *199*, 249–58, *250*, *251*, *252–3*, *255*, *256–7*

Becerra, Manuel 33
Bedford River, England 84–5, 86
Bedono, Java 279
Beymer-Farris, Betsy 202–3
Biesbosch National Park, Netherlands 154–5
Bilakoro, Diko 138
Bilanko, Victor 66
biologs 265–6
bio-rights 291
black-necked cranes *18–19*, 20, *21*, 24–5, 26
Bleuler, Johann Ludwig 150
blue carbon 272, 290
Bohai Sea, China 260, *261*, 265, *265*, 266
Boko Haram 128, 130
Bord na Móna 76
Boskalis 156
Boukary, Mana 130
Bozo ethnic group 133–4
'brave women of Kruscica' (protestor group) 182
Bruisch, Katja 75
Building with Nature Indonesia 283–4, 286–7
Bulbula, River, Ethiopia 56, 58
Bush, George H. W. 12
Buytaert, Wouter 33

C

Cacheu, River, Guinea-Bassau 248
Cácota, Colombia 31–2, 36
Camargue Wetland, France 168–9
Cantanhez National Forest, Guinea-Bassau 248
carbon emissions 7, 10, 33, 64–5, 70–1, 74, 75–7, 124, 190, 191, 194, 198, 268, 272–3, 274, 288, 290
Carter, Peter 84, *86–7*
Cassell, Stephen 237
Catsadorákis, Giórgos 173–4, 175, 179, 181
Center for International Forestry Research, Bogor, Indonesia 190
Central Rift Valley, Ethiopia 17, 49–60, *50*, *51*, *52–3*, *54–5*, *58–9*, 288
Chad, Lake 10, 80, *81*, 125–32, *126–7*, *129*, *130–1*
Chaffey, George 163

Charles I, King 85
Chernobyl Nuclear Power Station, Ukraine 75
Cheruthoni Dam, India 226
Chiaravalloti, Rafael Morais 106
chinampas (floating gardens) 36–8, *37*, 39
chinamperos (tillers) 37
climate change 6, 7, 10, 12, 33, 64, 70–1, 74, 75, 76, 90, 130, 184, 194, 256, 272, 273, 274, 289, 290, 291 *see also* carbon emissions
Cohen, Michael 164
Colorado River, U.S. 163–4, 218–19, 220
CORPONOR 32
Corredor Azul ('blue corridor') 110
Cortés, Hernán 36
Corumbá, Brazil 91, 97, 98, 109
Cromwell, Oliver 85
Cucapá people 218
Cuiabá, River, Brazil 98–100, *98–9*, 106, 109
Cuvette Centrale, Congo *161*, 189–95, *191*, *192–3*, *195*

D

Damasceno-Junior, Geraldo 101
Da Mosto, Jane 233
dams 9, 24, 32–3, 58, *58*, 59–60, 109, 110, 118, 122–4, 127, 128, 129, 130, 131–2, 138–9, 140, 148, 156, 165, 168, 170, 182, 185, 188, 190, *191*, 206, 211, 212, 217, 218–19, 220, 221, 226, 227, 228, 289, 290
Danube, River, Europe 11, 156, 220
Das, Saudamini 249
Dason, Siva 228
Dead Sea, The, Asia 185, 188
Debele, Abule 49, 50–1, 52, 53, 57
Débo, Lake, Mali 138
Delaware River, U.S. 265
Deltares 286
Demak, Java 276, *277*, *278–9*, 279, *282–3*, 283, 285, *285*, 286–7
depoldering 155, 260
desertification 20
Desta, Hayal 53
Dian, Lake, China 88, *89*, 90
Dinka herdsmen 142
Doñana Wetland, Spain 168

Donaye Taredji, Senegal 132
Dongting, Lake, China 90
dredging 12, 172, 188, 216, *217*, 229, 230, 232, 233, 236
Drozdova, Zoya 61–2, 64, 66–7
Dubna River, Russia 66
Duursche Waarden, River Ijssel, Netherlands 154
Duvail, Stéphanie 212
Dye, Barnaby 212
dykes 11, 12, 80, 90, 138, 147, 150, 152, *152*, 153, 154, 155, 156, 165, 217, 226, 228, 232, 236, 242, 248, 254, 259, 260, 267, 268, 275, 276

E
East Asian-Australasian Flyway 266
East Kolkata Wetland, India 168
Ebro, River, Spain 169, 220
Eden, The Garden of *8*, 9, 10, 12, 147, 148
embankments 70, 128, 254, 256, 257, 258
English Civil War (1642–51) 85
Erickson, Clark 38
Ethiopian Wildlife Conservation Authority 58
Etruscans 86
Euphrates River, Mesopotamia 147, 148
European Union 128, 130, 174, 175, 181
Everglades, Florida, U.S. 12, *196–7*, *198*, 267–74, *268*, *269*, *270–1*, *272–3*
Ezerani Wetland, Balkans 176

F
Faquibine, Lake, Mali 139
Felix, Tessa 116
Fens, the, England *81*, 83–5, *86–7*, 88
Fialho, Celma Francelino 105
flamingos 53, *54–5*, 56–7, 59, 169, 188
floodplains 11, 59–60, 64, 78–157, *78–9*, *80–1*, *82*, *86–7*, *89*, *92*, *93*, *94–5*, *96*, *97*, *98–9*, *102–3*, *108–9*, *113*, *114–15*, *118*, *119*, *120–1*, *126–7*, *129*, *130–1*, *134*, *135*, *136–7*, *138–9*, *143*, *144–5*, *146–7*, *151*, *152*, *153*, *154–5*, *157*, 188, 190, 212, 214, 228, 238, 288, 290
Flow Country, Scotland 75
Fomi Dam, River Niger 124, 140
frailejones (shrub) *14–15*, 16, 32, 33, *34–5*

Fredericks, Nicholas 111, 113, 116
Fryske Gea 259
Fulani people 127, 134, 139–40

G
Ganges, River, Asia 9, 11, 255
Gao Yufang 30
Ghadira Wetland reserve, Malta 169
Ghosh, Dhrubajyoti 165–6
Gibe III dam, Ethiopia *58–9*, 59–60, 124
Gle Jong 239–40
Global Mangrove Alliance 246, 248
Global Peatlands Initiative 76, 194
Gorgoram Fishing Festival 125–7
Gose, Ambila 222
Great Fen Project 88
Green Coast project 241
Greenpeace 194
Grill, Gunther 123
Grinchenko, Olga 66
Guadalquivir, River, Spain 168
Gu Haijun 24
Guinea-Bissau 248
Gumbricht, Thomas 190, 191

H
Hadejia-Nguru Wetland, Nigeria 125–9
Haigeng Wetland, China 90
Hamerlynck, Olivier 212
Hamis, Mohamed 208
Hamoun Wetland, Iran 185
Haramaya, Lake, Ethiopia 53
Helmand River, Afghanistan 185
Hergoualc'h, Kristell 191
Hidrovia canal, Paraguay 110
High Aswan Dam, Egypt 218
High Grand Falls Dam, Kenya 124
Hongyuan County, China 27–8
Hua, Lake, China 20, 24, 28
Humboldt Institute 33, 36
Hurricane Katrina 214, *215*, 216
Hussein, Saddam 147, 148
hydroelectricity 40, 48, 59, 109, 122, 124, 131, 140, 168, 182, 211, 212, 218, 220, 226, 288

I
Ibanez, Carles 220
Ijssel Meer, Netherlands 170, 172
Imperial Irrigation District, U.S. 164
Indonesian Ministry of Marine Affairs and Fisheries 283

Indonesian Ministry of Public Works and Housing 283
Indus, River, Asia 217–18
Inner Niger Delta, Mali *see* Niger Delta, Inner
International Fund for Saving the Aral Sea 184
International Peat Society 77
International Union for the Conservation of Nature 75, 206
Ireland 63, 75, 76–7, *76–7*
Irrawaddy, River, Myanmar 40
Isle de Jean Charles, U.S. 214
Ithai Barrage, India 40, 46, 48

J
Jaja, Indonesia 208–9, 211
James, E. J. 226
James, Tony 112
Jankulla, Vasil 178, 179
Jansen, Herco 53
Java, Indonesia *199*, 275–87, *276*, *277*, *278*, *280–1*, *282–3*, *284*, *285*, *286–7*, 291
Jeffries, Mike 170
Johnny, Angelbert 113
Joosten, Hans 20, 25

K
Kadiwéu people 106, 110
Kanchula, Kebede 57, 58
Kannady, India 222, 227
Karingamadathil, Manoj 223, 226
Kayrakkum Reservoir, Tajikistan 168
Kerala, India 12, *199*, 221–8, *223*, *224–5*, *226–7*
Keti Bunder, Pakistan 217
Keutajanga, India 250, 252, 254
KfW 64
Khamed, Nor 279
Khoiriyah, Sifatul 286
Khulna, Bangladesh 255–6, 258
Khwairamband market, Imphal, India 45
Kirk, Tom 164
Kirpotin, Sergei 70–1, 74
Kiyonga, Dia 210, 211
Kole Wetland, India 223–4, 227, 228
Kolkata, India 160, *161*, 165, 168
Kottayam Nature Society 222
Kramer, Johannes 260
Krueng Tunong, Indonesia 242–3
Kruscica, River, Balkans 182

Kumar, Ritesh 48, 226, 228, 254
Kuttanad Wetland, India 222, *223*

L
Laborde, Sarah 290
Laguna de Cácota, Colombia 32, 36
Lake Chad Basin Commission 130
land reclamation 222–3, 229, 230, 237, 250, 260
Lanzer, Toby 130
Laurinda de Carvalho, Maria 107
Lawson, Ian 190
Lenin, Vladimir 62, 63
Leopold, Aldo 218
Lethem, Guyana 114
levees 89, 213, 214, 257
Lewis, Simon 189–90
Liu Jian 27–8
loans, sustainable development project 241–2, 247, 284
Logone Wetlands, Cameroon 128, 290
Loitongbam, Babloo 46, 47
Loktak Development Authority 40, 48
Loktak, Lake, India 12, *17*, 39–48, *41*, *42–3*, *44*, *45*, *46–7*, 290
Loktak Protection Act (2006) 46, 48
Loon-Steensma, Jantsje van 260
Loth, Paul 128–9
Ly, Oumar Ciré 132
Ly, Seydou Ibrahima 132

M
Macaulay Institute, The, Scotland 107
Macedonian Ecological Society 177
Machão-Pão, Guyana 112–13
Macina Liberation Front 140
Mackenzie River, Canada 11
Ma'dan people (Marsh Arabs) 146, *146–7*, 147–8
Maga Dam, Chad 128
Magufuli, John 212
Mahamat, Aboukar 80
Mahanadi Delta, India 249–55, *250*, *251*, *252–3*, 291
Manantali Dam, Mali 131–2
Mangrove Capital Africa 206
mangrove kingfisher 208, *209*

mangroves 9, 11, 13, 62, 169, *196–7*, 198, *200*, 201, 202, 203, *203*, 206, 207, 208, *209*, 210, 211, 212, 218, 222, 223, 237, 238, 240, 241–3, *242*, *244–5*, 246–8, *246–7*, 249–50, *252–3*, 254, 255, *255*, 256, *256–7*, 257, 258, 260, 267, 268, 273, 274, 275, 276, 279, 282, 284, 285, *285*, 286, 287, 290
Manhattan, U.S. 237
Manipur, India 40, 45, 46, 48, 288
Mao Zedong 20, 90
Marker Wadden, Netherlands 170, *170*, 172
Masrur, Muhammad 279
Maximo da Silva, Jacinto 101, 104
Mecklenburg-Vorpommern, Germany 76
Medvedev, Dmitry 64
Meitei people 12, 40, *41*, 44, *44*, 45–8
Meki Batu Fruit and Vegetable Growers Cooperative Union 57
Mekong, River, Asia 118, 119–20, 122–3, 124
Mekonnen, Solomon 56
Mequinenza Hydroelectric Dam, Ebro 220
Merkel, Angela 64, 153
Meschera National Park, Russia 61–2, 64, 66, *66–7*
Mesopotamian Marshes, Iraq 9, 12, 80, *81*, 141–8, *143*, *144–5*, *146–7*, 185
Mexico City 37–8
migration, human 127, 128, 129–30, 140, 185
Milky storks *284*, 285
Minayeva, Tatiana 68
Miranda, River, Brazil 93, 96, 97, *97*
Mississippi Delta, U.S. 89, *198*, 213–17, *215*, *216*, *217*, *218–19*
Mississippi River Commission 213
Mondal, Shahjahan 257–8
Morelos Dam, U.S. 219
Mörschel, Frank 64
Moscow Peatlands, Russia *17*, 61–8, *62–3*, *65*, *66–7*
Mousini Island, Sundarbans 256, *256–7*
mudflats 11, 214, 232, 237, 260, *261*, 264, 266, 283
Mulonga, Julie 206
Mulugeta, Amdemichael 57, 58
Murdiyarso, Daniel 273, 274
Mwalyosi, Raphael 211–12

N
Naivasha, Lake, Kenya 59
Nakuru, Lake, Kenya 56
National Mangrove Ecosystem Management Strategy 274
National Wetlands Research Institute 110
Natron, Lake, Tanzania 59
Nature Iraq 148
Ndiaye, Amadou Lamine 132
neotropic cormorant *92*, 93
New Bedford River, England 85, 86
New Orleans, U.S. 213, 214, *215*, 216
Nicola, Rafaela 100, 106, 110
Nieuwe Merwede, Netherlands 154, 155–6
Niger Delta, Inner, Nigeria *2–3*, 4, 9, 10, 80, *81*, 124, 133–40, *134*, *135*, *136–7*, *138–9*, 190, 290
Nile, River, Egypt 9, 10, 60, 141, *143–4*, 183, 202, 217, 218, 288
Niño, Sergio Ivan 32
Noard – Fryslân Bûtendyks, Netherlands 259, 260–1, *262–3*, 264
Noga, Mali 135, 138
Noor, Yus Rusila 283
Norfolk Broads, England 170, 171, *172*
Noyabrsk, Russia 70
Ntilicha, Mathew 206, 207
Nunes da Cunha, Cátia 104

O
O'Hanlon, Redmond 189
Ohrid, Lake, Balkans *161*, 179, 181
Okavango Delta, Botswana 12, 288
Okeechobee, Lake, U.S. 268
Omo, River, Ethiopia 59, 60, 124
Orford, Julian 258
Orinoco River, South America 33, 190
Orshinsky Bog 65–6
Ouse, River, England 84
Oxfam Novib 241

P
Pacific Institute 164
Padma, River, Asia 255, 256
palm oil plantations 76, 190, 191–2, *192–3*, 194, 246
Panama City 238
Panov, Vladimir 65–6
Pantanal, Brazil 9, 11, 80, *80*, 91–110, *92*, *93*, *94–5*, *96*, *97*, *98–9*, *102–3*, *104–5*, *108–9*, 288, 289

pantaneiros (indigenous people) 98, 100, 101, *102–3*, 104, 107, 110
Papandreou, George 175
Paraguay, River 11, 97, 98, 109, 110
Parameswaran, K. 228
Paramo De Suntarban, Colombia *14–15*, 16, *17*, 31–8, *34–5*, 37
Paraná, River, Brazil 11, 109, 110
Parime, Lake, South America 112
Paris climate agreement (2016) 76, 274
Passo do Lontra, Brazil 93–4, 97
Peatland Restoration Agency 194
peat bogs/peatlands 10, 20, 21, 24, 29, 61–8, *62–3*, 69–77, *70–1*, *72–3*, *76–7*, 83, 170, 179, 181, 189–91, *191*, 194, *195*, 267, 268, 272, 273, 290
PeatRus (Russian Peatlands Restoration Project) 64
pelicans 117, 138, 164, 173–5, *175*, 177, 179, *218–19*, 226
Periyar River, India 226–7
permafrost 9, 16, 69–77, *70–1*, *72–3*, *76–7*
permeable pavements 237, 238
Pethick, John 258
Philippi Bog, Greece 75
phumdis (floating islands) 44–8, *46–7*
pink-plumed greater flamingos 169
pink-plumed lesser flamingos *54–5*, 56
Poconé, Brazil 101, 104, 106, 107, 109
polders (dyked areas of land designed to keep out floodwaters) 154, 155–6, 223, 256, 259, 260, 261
pollution 10, 53, 64, 70, *71*, 90, 109, 163, 222, 226, 228, 246, 266, 288
Poyang, Lake, Chia 188
Prespa, Lake, Balkans 12, 160, *161*, 173–9, *175*, *176–7*, *180*, 181, 182, 290
Pripyat peatland 75
Protection and Preservation of Natural Environment in Albania 178

Q

Quechua people 38

R

Radchenko, Ivan 62–3
Ramsar Convention on Wetlands, The 10, 21, 106, 168, 176, 228, 238
Rashid, Kassim 208
rectification, river 149–53, 214

reservoirs 59, 109, 119, 168, 211, 218, 220, 226, 238
Responsibly Produced Peat Foundation 77
rewetting 24, 62, 64, 65–6, 67, 76–7, 88, 264, 268
Rhine, River, Europe 11, *81*, 128, 149–57, *151*, *152*, *153*, *154–5*, 156, *157*, 214, 260, 290
Rhône, River, Europe 165, 169, 220
Ribeiro, Antonia and Leandro 98–9
Rift Valley Lakes Basin Authority 57, 58
Rockwood, Charles 163
Rodrigues, Altemisia Dias 101
Roman Empire 83–4, 86, 229
Roque, Fabio 96, 109
rose farms 51, 52–3, *52*
Roucoux, Katy 190
Rouhani, Hassan 188
Rufiji Basin Development Authority 211
Rufiji Delta, Tanzania 124, *199*, *200*, 201–12, *202–3*, *204–5*, *209*, *210–11*, 290
Rufiji Environment Management Project 203, 206
Ruiz, Daniel 33
Ruma, Tanzania 208
Ruoergai Plateau (Zoige Plateau), China *17*, *18*, 19–30, *21*, *22–3*, *26–7*, *29*, 290
Rupununi, Guyana *78–9*, 80, *80*, 111–16, *113*, *114–15*
Rutland Water 168

S

Sabino, José 91
Saemangeum Estuary., South Korea 266
Sahara Desert, Africa 4, 9, 124, 130, 138
Sahel, Africa 10, 124, 128, 130, 131
Sairi, Mat 276, 283
Salelie, Yusuph 207
Salim, Agus 279
salinas (industrial salt pans) 168–9
Salin de Giraud, France 168–9
salt marshes 165, 198, 260–1, *261* *262–3*, 264, 265
Salton Sea Authority 164
Salton Sea, California, U.S. 160, *160*, *162*, 163–5, *165*, *166–7*
Sanajaoba, Ngangom 40, 44, 45, 48
Sanankoua, Daouda 9

San Diego, U.S. 164
San Francisco Bay, U.S. 238
sangai (deer) 39–40, 44, 45
São Pedro de Joselândia, Brazil 100–1, 104, 106–7
Sarmiento, Carlos 36
Schikker, Arno 156
sea level rise 36, 232, 246, 258, 260, 264–5, 268, 286
seasonal floods 60, 93, 123, 131, 132, 148
sea walls 237, 254, 260, 264, 266
Semarang, Java 276, *276*, 286
Sembilang National Park, Indonesia 247
Senegal, River 131–2
Senegal River Basin Development Organization 132
sewage 160, 165, 168, 246
Shapiro, Judith: *Mao's War Against Nature* 90
Shepard, Christine 264
Sher Ethiopie 52, 53, 57, 58–9
Shulinab, Guyana 111
Siberia 9, 62, 69–77, *70–1*, *72–3*, *76–7*, 190, 272
Sichuan Ruoergai Wetland National Nature Reserve, China 24
Sichuan Wetlands Management Center, China 24
Simina village, Mali 138
Sindh Desert, Pakistan 217
Sistani people 185
Smithsonian Institution, Washington D.C. 6
Sobha City, India 223
Social Service for Commerce (SESC) 106, 107
Society for the Protection of Prespa 174
soda ash 51, 56–7, 58–9
Somerset Levels 88
Sonoran Desert, North America 218
SOS Ohrid 181
Staatsbosbeheer 155
Stankovic, Sinisha 181; *The Balkan Lake Ohrid and its Living World* 181
Stanley, Henry Morton 189
St Elizabeth's Flood (1421) 154
Stenje Marsh, Macedonia 177
Stiegler's Gorge Hydroelectric Dam, Tanzania 211–12
Stive, Marcel 258

Studenchishte Marsh, Macedonia 179–80, 181
Sudd swamp, South Sudan 10, *81*, 141, 142, *143*, *144–5*, 146, 190, 288
Sundarbans, Bay of Bengal 11, 250, *252–3*, 255, *255*, 256, *256–7*
Supercyclone Kalinga 249–50, *251*, *252–3*, 254
superstorm Sandy 237
Suryadiputra, Nyoman 242
Swain, Pramod 250, 254
Syr Darya River, Central Asia 183, 184

T

Taldom Homeland Nature Reserve, Russia 66
Talevski, Trajce 181, *182*
Tana, River, Kenya 124
Tandahar, Indonesia 250–1, 254
Tanzania Forest Service 206
Taquari River, Brazil 98
Tasin, Russia 68
Tenochtitlan, Mexico 36
Terena people 105
Texcoco, Lake, Mexico 36
Thames River, England 84, 170, 238, 264
Thijssen, Sylvo 155
Thijssen, Wil 155
Thorne, Colin 25
Tibet 9, 16, *18*, 19, 20, 24, 25, 26, 27, 29, 30, 118, 217
tidal surges 198, 214–15, 255, 256, 257, 258, 264
Tigris river, Mesopotamia 147, 148
Timbuktu, Mali 9, 133, 134, 139
Timbulsloko, Indonesia 275–6, 278–9, 282, 283, 286, 287
Titicaca, Lake, Peru 12, 38
Tonle Sap, Cambodia 80, *81*, 117–24, *118*, *119*, *120–1*
tourism 12, 24, 25, 26, 28, 32, 36, 37–8, 40, 44, 53, 67, 97, 98, 100, 101, 104, 106, *108–9*, 109, 110, 116, 119, 132, 172, 181, 182, 188, 221, 222, *224–5*, 226, 228, 229, 230, 232, *233*, 236, 285, *285*, 288
Trajanovski, Vladimir 181
Trajce, Aleksander 178, 179
Transpantaneira, Brazil 108–9
Traoré, Yousseff 138
Trittin, Jürgen 156
Troulet, Baba 135

tsunami, Indian Ocean (2004) 239–42, *240–1*, *242*, *244–5*, 245, 246, *246–7*
Tulla, Johann 149–50, 151, 156
Tuminec, Albania 179
Turkana, Lake, Kenya 59, 60, 124
Tver, Russia *62–3*, 65–6, 68
Twain, Mark 214

U

Ubangi, River, Congo 130
UNESCO 147
United Nations 142, 184, 188
 Development Programme 188
 Environment Programme 128, 147
 Food and Agriculture Organization 122
 International Organization for Migration 130
 Intergovernmental Panel on Climate Change 273
Urmia, Lake, Iran 188
Ushe, Nugusa 53

V

Valeriy, Ignatiev 68
Varela, Juan Carlos 238
Veldkamp, Ted 124
Vembanad, Lake, India 222, 223, 226
Venice, Italy 36, *199*, 229–37, *230*, *231*, *233*, *234–5*, *236–7*
Vermuyden, Cornelius 84, 85, 88
Vicente da Silva, Luiz 107, 108–9
Vierlingh, Andries 154
Volga, River, Russia 65, 66

W

Wadden Sea, Netherlands 170, *199*, 259–64, *261*, *262–3*, 283
Wallasea Island, Essex 264–5
Wapichan people 12, 80, 111, 112, 113, *113*, 116
Waqie Wetland 28
Warufiji people 203, 207, 211
Waza National Park, Cameroon 128
Wedung, Java 279, 284, 285
West Kalimantan, Borneo 190, *191*
West Thurrock Lagoon, Essex 170
Western Ghats, India 221–2, 226
Western lowland gorillas 194, *195*

Wetlands International 6, 24, 48, 57, 64, 68, 70, 76, 77, 100, 106, 109, 110, 138, 140, 147, 190, 192, 194, 206, 211, 223, 226, 238, 241, 242, 248, 254, 266, 279, 283, 285, 290–1, 303, 304
Widodo, Joko 194
World Bank 258
World Wetlands Day (2018) 237–8
WWF (World Wildlife Fund) 206, 212

X

Xayaburi Dam, Laos 122
Xiangdong, China 26
Xiaowan Dam, China 122

Y

Yangtze River, China 20, 90, 188
Yellow River, China 16, 20, 21, 24, 30
Yellow Sea, Asia 260, 264, 266

Z

Zaec, Daniela 177
Zenawi, Meles 124
Ziway, Lake, Ethiopia 49, 51, *51*, 52–4, 56, 57, 59
Zwarts, Leo 139, 140

ACKNOWLEDGEMENTS

Many people helped this book with their observations, enthusiasm and analysis. Invaluable technical and practical guidance was given on the whole project by Roz Kidman Cox. From Wetlands International's global office we thank Arno Kangeri, Mira-Bai Simón and Gina Lovett for their parts in bringing the concept and contents together. A lot of those we met during our researches are acknowledged by being quoted in the text, starting with Daouda Sanankoua in the Inner Niger Delta, and ending with Sifatul Khoiriyah in the flooded Javanese village of Timbulsloko. But many other people helped, especially with the logistics of arranging the journeys that make up the majority of the book. We cannot name them all, but our special thanks are due to the following people.

In China, on the Ruoergai Plateau, our guide and mentor was Xiaohong Zhang of our China office. Thanks also to Erga and especially Yong-Xiu of the Ruoergai National Nature Reserve, Gu Haijun of the Sichuan Wetlands Management Center in Chengdu, and local officials in Hongyuan and Ruoergai counties. Our work in India was masterminded by Ritesh Kumar, Wetland International's South Asia Director, and organized on the ground by the resourceful Dushyant Mohile. At Lake Loktak we were hosted by the Loktak Development Authority, whose Sanajaoba Ida was a candid guide. On the Mahanadi Delta, Pranati Patnaik for Partners for Resilience speeded our journey.

In Ethiopia, Wetland International's Amdemichael Mulugeta organized our journey to the lakes of the Central Rift Valley in the company of Julie Mulonga, our Eastern African Director. We were guided at Lake Ziway by Nugusa Ushe. Thanks also to Kebede Kanchula, Director-General of the Rift Valley Lakes Basin Authority, and the Meki Batu Fruit and Vegetable Growers Cooperative Union. In Russia, our associate Tatiana Minayeva organized and guided us. Thanks also to Zoya Drozdova at Meschera National Park, Vladimir Panov and others at the Tver State Technical University, as well as officials at Kameshkovo and the Taldom Homeland Nature Reserve.

The marvels of the Pantanal in Brazil were revealed to us by Rafaela Nicola and her staff at our Brazil office in Campo Grande, along with field researchers Fabio Roque, Cátia Nunes da Cunha and Jose Sabino, rancher Luiz Vicente da Silva, the butterfly farmers of Poconé, and the people of São Pedro de Joselândia. Thanks also for the hospitality of the Hotel Sesc Porto Cercado. We journeyed to the former floodplain of the River Senegal with Papa Diomaye Thiare of our Africa office in Dakar, and to the Inner Niger Delta with its estimable former head in Mali, Bakary Kone.

In the Netherlands, we visited Noordwaard and Marker Wadden with Arno Schikker and Frans Uelman of construction company Boskalis, and restored salt marshes with Chris Bakker of It Fryske Gea. We learned the delights of the multinational Lake Prespa in Greece with Julia Henderson and Giórgos Catsadorákis, the inspirational founder of the Society for the Protection of Prespa; in North Macedonia with Daniela Zaec of the Macedonian Ecological Society; and in Albania with Aleksander Trajce, who runs Protection and Preservation of Natural Environment. And we toured the shores of Lake Ohrid with Vladimir Trajanovski of Ohrid SOS. Thanks also to Daniel Scarry.

The villages of the Rufiji Delta in Tanzania were introduced to us by the Tanzania Forest Service's Mathew Ntilicha, in the company of Julie Mulonga. Thanks also to the village committees in Ruma, Nyamisati, Mfisini and Jaja. In Indonesia, we visited the tsunami survivors in Aceh with our Indonesia Director Nyoman Suryadiputra, along with Kuswantoro. And we toured Demak district in Java with Yus Noor and Apri Susanto Astra from our Indonesia office. Thanks also for the hospitality in the besieged villages of Timbulsloko, Tugu, Wedung, and Bedono.

FP would like to thank others for help with other reporting which has been incorporated into this text. His visit to the Páramo de Suntarban in Colombia was facilitated by SABMiller and guided by Sergio Ivan Niño of the local conservation authority COPRONOR. FP went to Noyabrsk with Sten Nilsson of the International Institute for Applied Systems Analysis; to Guyana with Tom Griffiths of the Forest Peoples' Programme; to the Tonle Sap in the company of WorldFish's Eric Baran; to Nigeria's Hadejia-Nguru Wetland on commission for *Audubon* magazine; and to the Aral Sea in Uzbekistan first with UNDP and later with the International Water Management Institute.

CREDITS

Getty Images: Back cover, 4, 5, 14-15, 18, 37, 42-43, 71, 86-87, 126, 131, 155, 158-159, 186-187, 215, 217, 233, 244-245, 252-253, 255. Alamy Stock Photo: Back cover, 4, 34-35, 65, 72-73, 78-79, 118, 119, 120-121, 129, 136-137, 143, 171, 200, 210, 216, 243, 256, 269. National Geographic Image Collection: Front cover, David Doubilet; back cover, 2-3, 5, 41, 54-55, 165, 166-167, 196-197, 204-205, 209, 219, 234-235, 268, 270-271. Nature Picture Library: Endpapers, 27, 82, 162, 224-225, 272.Greenpeace: 89, 191, 192-193, 265. Minden Pictures: 114-115, 157. Wetlands International: 46-47, 251, 257, Dushyant Mohil; 51, Amdemichael Mulugeta; 135, Jane Madgwick; 138-139, Ibrahima Sadio Fofana; 278, 284, Yus Rusila Noor; 202-203, Joy Kivata; 241, I. Nyoman Suryadiputra; 242, Pieter van Eijk; 286-287, Yoyok Wibsono. Society for the Protection of Prespa: 175, 176-177, Thanos Kastritis; 178; Christina Ninou; 180, Andrea Bonetti. Nature Iraq: 146-147, Jassim Al-Asadi. Royal Boskalis Westminster N.V.: 170. Beeldbank Rijkswaterstaat: 262-263, Joop van Houdt; 8, Creative Commons; 21, Xuesong Han; 22-23, Yuriy Rzhemovskiy on Unsplash; 29, Gu Haijun; 44, 226-227, 246-247, Fred Pearce; 45, Amit Bansal Images; 50, Ninara; 52, Adriano Marzi; 58, Fausto Podavini; 63, 66-67, Roman Kakotkin, 77, Dennis Horgan; 92, 93, 94-95, 97, 98-99, 102-103, 104-105, 108-109, José Sabino; 96, Rafael Hoogesteijn - Panthera Brasil; 113, Forest Peoples Programme; 134, Leo Zwarts; 144-145 & 304, Yann Arthus-Bertrand; 151, © Liechtensteinisches Landesmuseum (Foto: Sven Beham); artist: Johann Ludwig Bleuler (1792-1850). 152, ©Lebendiger Alpenrhein 2011; 153, ©Peter Rey / Lebendiger Alpenrhein 2011. 184-185 & 231, Images Courtesy of NASA; 195, Will Burrard-Lucas; 223, REUTERS/Sivaram V; 230, Library of Congress, Geography and Map Division; 236, Adam Batterbee; 261, Jan van de Kam; 276, 282, 285, Cynthia Bol; 277, Tom Wilms; 280-281, Nanang Sujana.

COVER: A fisherman creates a silhouette from his canoe over a lily forest in the Okavango Delta of Botswana.

ENDPAPERS: A coot swims past reed beds, as the sun rises over a misty Lakenheath Fen in Suffolk, England.

FOLLOWING PAGE: Everyone loves mangroves. But they rely on a mix of fresh and salt water to thrive. On this island in New Caledonia, where sea water only penetrates during spring tides, nature has drawn this glade amongst the mangroves in the shape of a heart.

William Collins
An imprint of HarperCollins*Publishers*
1 London Bridge Street
London SE1 9GF
WilliamCollinsBooks.com

First published by William Collins in 2020

Text © Wetlands International 2020
Photographs © individual copyright holders
Cover and interiors designed by David Griffin
Picture research by Caroline Cortizo at Shifting Pixels

10 9 8 7 6 5 4 3 2 1

Wetlands International assert their moral right to be identified as the authors of this work

All rights reserved. No parts of this publication may be reproduced, stored in a retrieval system or transmitted, in any form or by any means, electronic, mechanical, photocopying, recording or otherwise, without the prior permission of the publishers.

A catalogue record for this book is available from the British Library.

Trade ISBN: 978-0-00-839049-5
Special edition ISBN: 978-0-00-838413-5

All reasonable efforts have been made by the authors and publishers to trace the copyright owners of the material quoted in this book and any images reproduced in this book. In the event that the authors or publishers are notified of any mistakes or omissions by copyright owners after publication, the authors and publishers will endeavor to rectify the position accordingly for any subsequent printing.

Printed and bound in China by RRD Asia Print Solutions

MIX
Paper from responsible sources
FSC™ C007454

This book is produced from independently certified FSC™ paper to ensure responsible forest management.

For more information visit: www.harpercollins.co.uk/green